A Brief History of High-Speed Rail

Qizhou Hu · Siyuan Qu

A Brief History
of High-Speed Rail

Qizhou Hu
School of Automation
Nanjing University of Science
and Technology
Nanjing, Jiangsu, China

Siyuan Qu
China Railway Shanghai Group Co., Ltd.
Shanghai, China

ISBN 978-981-19-3637-1   ISBN 978-981-19-3635-7 (eBook)
https://doi.org/10.1007/978-981-19-3635-7

Jointly published with Southwest Jiaotong University Press
The print edition is not for sale in China (Mainland). Customers from China (Mainland) please order the
print book from: Southwest Jiaotong University Press
ISBN of the Co-Publisher's edition: 978-7-5643-6259-1

This Springer imprint is published by the registered company Springer Nature Singapore Pte Ltd.
The registered company address is: 152 Beach Road, #21-01/04 Gateway East, Singapore 189721,
Singapore

Preface

As the saying goes, "the height you stand decides your view, the points of view you observe changes your concept, and the measure you choose grasps a better life." While "speed" determines transportation, "transport capacity" changes the transport modes and the "technology" ensures the traffic safety. High-speed rail (HSR) is regarded as a vehicle that changes human life and characterized by "There is not a fastest vehicle, only faster." Compared with other transport modes, HSR has many advantages such as strong transportation capacity, high operation speed, high safety, high punctuality rate, low-energy consumption, little influence on the environment, land conservation, comfort, convenience, considerable economic and social benefits. With its unique technical advantages, HSR has adapted to the new demands for economic development of modern social. As a result, HSR has become an inevitable choice for the development of countries around the world. In particular, the development and operation of China's high-speed rail indicate that HSR has great development space and potential in China. China would better make full use of its latecomer advantage to realize the leap-forward development of HSR. Therefore, *A Brief History of High-Speed Rail* is a reading book that introduces the basic knowledge, concept terms and development history of HSR and sorts out the main results of HSR at this stage. This book mainly explains the connotation of HSR for readers from two different aspects of theories and technologies. Combining the characteristics of high-speed rail and the development trend of the world, this book introduces the development meaning of high-speed rail. The main contents are as follows:

(1) Concept term of HSR. Related terms of HSR, especially the definition of HSR and the speed classification, as well as the main attributes of HSR, such as speed, capacity, safety, comfort, economy, energy efficiency, environmental protection and so on.

(2) Development trend of HSR. Introducing the development trend, technological features and application prospect of the three classes of Wheel High-speed Rail (WHSR), Magnetic High-speed Rail (MHSR) and Super-speed Rail (SSR), those are the past, present and future of HSR development. This book is mainly about two aspects. On the one hand, the author describes the development of

HSR in time—yesterday, today and tomorrow. On the other hand, the author comparatively analyzes the development trend of HSR in domestic and foreign countries (especially the HSR comparative analysis of Japan, France, Germany and China) in space. Through the comparative analysis, readers can understand the development trend of China's HSR in the world, ranking first in terms of "quality" or ranking first in "quantity."

(3)   Regional integration in the environment of HSR. Various types of HSR promote integration level among different regions. WHSR promotes the regional integration, MHSR promotes the continental integration, and SSR promotes the world's integration, leading to a global village. Therefore, we must pay attention to the development of HSR. At present, countries that master the core technology of WHSR, MHSR, SSR, etc., will own the world.

This book is created by the Qizhou Hu team of High-Speed Rail Science Research Institute. The team members mainly include Senior Engineer of Shanghai Railway Bureau Siyuan Qu, graduate students of the Nanjing University of Science and Technology Jie Chen, Ziquan Cong, Minjia Tan, Airan Zeng, Lishuang Bian, Xiaohan Li, Min Yue, Yikai Wu, Xin Guan, Song Ding, Longxin Zheng and Xiang Lin.

For a popular science book, benefiting the public is the highest aim. However, some of the pictures and contents of this book come from the Internet. Since we cannot find the source, we can only express our gratitude and respect here for them. We are also grateful for the selfless help from the editorial staff when we are in the writing of this book.

This book can be used as a reading material for researchers, engineers and technical personnel, management workers, college teachers and students and high-speed rail enthusiasts. Limited by time and knowledge of us, there are inevitable omissions and inadequacies in the book. Please enlighten us. Thank you.

Nanjing, China   Qizhou Hu
Shanghai, China   Siyuan Qu
August 2018   The Writer

Contents

Chapter 1
Introduction

With the official operation of the world's first High-speed Rail (HSR) in Japan on October 1, 1964, HSR opened a new era of transportation development. "There is not a fastest vehicle, but only faster vehicle", speed and capacity are eternal pursuit of mankind. No matter what kind of transportation (trains, cars, airplanes, ships, etc.), the requirements of human beings for it not only depend on speed but also on transport capacity.

Although the aircraft operates at a high speed, its transport capacity is limited. While the train has a large transport capacity, but it runs at a low speed. Therefore, the pace of human's pursuit of transport never ceased, and HSR is the crystallization of human wisdom in transportation. The French Wheel High-speed Rail TGV (train à grande vitesse, TGV) is as shown in Fig. 1.1.

HSR is an abbreviation for high-speed rail. It is a large system composed of dedicated lines, high-speed trains, and dedicated control systems. Therefore, HSR is a system concept but not an individual concept. The "high-speed" in the high-speed rail refers to the quality, while the "rail" is the property. In addition to Wheel High-speed Rail (WHSR), HSR also includes Magnetic High-speed Rail (MHSR) and Super-speed Rail (SSR). Therefore, the narrow concept of HSR refers to the WHSR transport system. The broad concept of HSR includes not only the WHSR transport system, but also the MHSR, which is using the magnetic levitation technology, and the SSR transport system in the vacuum track. Figure 1.2 is a diagram of high-speed rail train.

HSR has become a hot issue in the world. This is because HSR has some technological advantages that are incomparable to other modes of transportation. The first advantage is the high speed. The French WHSR TGV set a world record with the speed of 574.8 km/h. Japan's MHSR set a world record with the speed of 603 km/h. The America's SSR set a world record with the speed of 1000 km/h, and it faster than the normal speed of the airplane. The normal speed is 800 km/h. The second advantage is the large volume. The interval of HSR trains can be as short as 4 min and twelve trains can be operated per hour in one direction, which is incomparable to

© Southwest Jiaotong University Press 2023
Q. Hu and S. Qu, *A Brief History of High-Speed Rail*,
https://doi.org/10.1007/978-981-19-3635-7_1

Fig. 1.1 French WHSR: TGV

Fig. 1.2 HSR train

highways and aviation. The third advantage is the high safety. The quality and precision of HSR line facilities are high. The train operation control system uses mature electronic technology and intelligent software which ensures the safety distance between the two trains. Therefore, there are few accidents in HSR around the world. Fourthly, HSR can operate throughout the day because it cannot be affected by rain, snow, fog, wind. Fifthly, HSR also has the features of low energy consumption, land conservation, light pollution and high comfort. Therefore, HSR has been welcomed by most countries in the world since its birth.

1.1   Emerging Conditions of HSR

(1)   The production of vehicle requires certain conditions. No matter which kind of transportation, human beings appraise it from three aspects. The vehicle that can meet these requirements is good, but not vice versa. First is the functionality such as speed, capacity, and safety. Second is the economics such as cost, energy, and efficiency. The final one is the ecology such as noise, radiation, and environmental protection. As a means of transportation, HSR also takes the load into account while pursuing high speed. Noting that high speed and heavy loads are the eternal pursuit of mankind, HSR exactly meets human needs. Figure 1.3 shows Wheel High-speed Rail system.

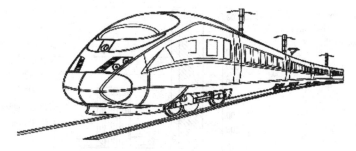

Fig. 1.3 WHSR system

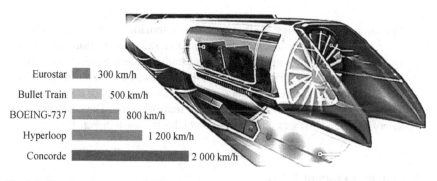

Fig. 1.4 The operating speed of different vehicles

(2) The speed of HSR. Speed is the basic requirement for transportation. It is exactly the rapidity and high efficiency that make the HSR popular with humans and developed greatly. A comparison of speed between HSR and other vehicles is shown in Fig. 1.4.

(3) The load of HSR. In order to meet the basic demand for transportation, people expect vehicles to carry as much weight as possible. The comparison of the loading capacity between HSR and existing vehicles (cars, airplanes, ships, traditional trains, etc.) is shown in Fig. 1.5, from which we can obtain that the HSR is the vehicle with the largest load.

Fig. 1.5 Comparison of the loading capacity between HSR and other vehicles

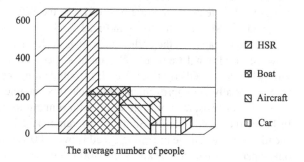

Fig. 1.6 Comparison of safety between HSR and other various vehicles

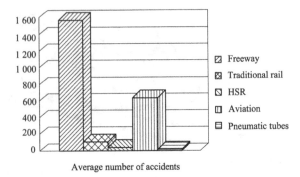

Average number of accidents

(4)   The safety of HSR. Since HSR is operating automatically in a fully enclosed environment and has a series of comprehensive safety protection systems, its safety is unmatched by any other means of transportation. Several major HSR countries have to operate thousands of HSR trains every day. While the accident rate and casualty rate are far lower than other modern modes of transportation. Therefore, HSR is considered as the safest transportation. The comparison of safety between HSR and other various vehicles is shown in Fig. 1.6.

1.2   Three Leaps of HSR

In order to satisfy the demands for both speed and capacity, the HSR experienced three qualitative changes, namely three leaps. From WHSR to MHSR to SSR, from the operating speed of 200 to 500 to 1200 km/h, this is also the three leaps in human demand for transportation.

1.2.1   The First Leap: Improve the Speed of Operation and the Birth of WHSR

The first category: Wheel High-speed Rail (WHSR). To improve the speed of the train brings the first leap. So the first type of HSR, the WHSR was born. The traditional train and WHSR train are shown in Fig. 1.7.

In terms of capacity, the traditional train is the king of all modes of transportation. However, traditional trains usually run at a speed below 200 km/h, which cannot satisfy human's needs for fast travel. By strengthening the study of track and vehicle type, especially the improvement of vehicle type, people reduce the frictional resistance and air resistance of high-speed train running to increase the running speed. In Japan, the speed of HSR (Shinkansen) train reached 200 km/h in 1964. The train is called HSR when the operating speed is over 200 km/h. However, the WHSR can only operate between 200 km/h and 400 km/h due to air resistance and frictional

Fig. 1.7 Traditional train and WHSR train

resistance. The operating speed of 400 km/h is the warning threshold of WHSR. When the WHSR train exceed this speed, it is extremely easy to derail and cause traffic accidents. WHSR train is as shown in Fig. 1.8.

Wheel High-speed Rail is mainly a transportation system running on the track, which is generally shortened for WHSR and can also be called conventional HSR. The main features are as follows:

① The operating speed of the WHSR is about 200–400 km/h.
② The warning threshold of WHSR is 400 km/h.
③ The resistances of WHSR are frictional resistance and air resistance.

WHSR belongs to the wheel-rail type of HSR. According to the definition of the International Railway Union, HSR refers to the railway system that has an operating speed of more than 200 km/h by transforming the traditional line (straight line, gauge standardization), or has an operating speed of more than 250 km/h by building a new line. This book divides WHSR into three types. See Table 1.1 for details.

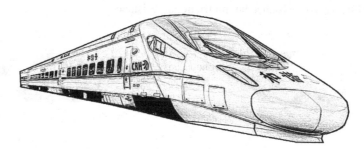

Fig. 1.8 WHSR train

Table 1.1 Types of WHSR

Number	Types	Speed/(km/h)	Name	Main countries	Remarks
1.	First	200–300	Low-speed WHSR	Japan, Germany	The warning threshold of WHSR is 400 km/h
2.	Second	300–350	Normal-speed WHSR	France, China	
3.	Third	350–400	High-speed WHSR	China	

1.2.2   The Second Leap: The Removal of Frictional Resistance Brings the Birth of MHSR

The second category: Magnetic High-speed Rail (MHSR). In order to reduce the friction between the wheels and the rails, the second leap was made. As a result, MHSR, the second type of HSR, was born. The WHSR and MHSR trains are as shown in Fig. 1.9.

In order to reduce the frictional resistance and improve the running speed and meet the fast travel requirements of human beings, MHSR was born with the running speed of more than 400 km/h based on the principle of "same-magnet repelling and opposite-magnet attraction". During the operation of the MHSR, the magnet train does not directly contact the track, but floats on the track so that there is no frictional resistance and then the running speed is improved. In 2015, the speed of MHSR in Japan has reached 600 km/h and more. Although MHSR is not affected by the frictional resistance, it can only operate at the speed of 400–1000 km/h due to the limitation of air resistance. The operating speed of 1000 km/h is the warning threshold of the MHSR. When this speed is exceeded, the operating cost will be too high. Among them, Japan's MHSR train is as shown in Fig. 1.10.

MHSR is the magnetic suspension type of HSR, which is mainly suspended on rails to run. It is also called superconducting high-speed rail. The main features are as follows:

① The operating speed of MHSR is from 400 km/h to 1000 km/h.
② The warning threshold of MHSR is 1000 km/h.
③ MHSR has air resistance but no frictional resistance.

Fig. 1.9 WHSR and MHSR trains

Fig. 1.10 Japan's MHSR train

Table 1.2 Types of MHSR

Number	Types	Speed/(km/h)	Name	Main countries	Remarks (0 K = −273.15 °C)
1.	First	400–600	Low-temperature MHSR	Japan, Germany	4.2 K—Liquid helium (rare, expensive)
2.	Second	600–800	Normal-temperature MHSR	Japan	15 K—Liquid helium (minor, reasonable)
3.	Third	800–1000	High-temperature MHSR	Japan	77 K—Liquid helium (much, cheap)

MHSR belongs to the magnetic suspension of HSR. As a new type of ground transportation, magnetic train has moved from the experimental stage to commercial operation and overcome the problems of traditional train such as the adhesion limit, mechanical noise and wear, etc. Besides that, MHSR has the features of high speed, strong climbing ability and low energy consumption, high noise, high safety, high comfort, no fuel, and little electromagnetic pollution. It has become the ideal vehicle for people.

The MHSR train can be divided into two types based on the principle of suspension: Electromagnetic Suspension (EMS) and Electrodynamic Suspension (EDS). The speed of MHSR train can reach 500 km/h, which is absolutely impossible for traditional train. If the superconducting magnet is installed in the train and an aluminum ring is laid on the ground track, the relative movement between them will generate an induced current in the aluminum ring. Then the magnetic repulsion will occur, lifting the train about 10 cm from the ground, allowing the train to float on the rail and operate at a high speed. This book divides the MHSR into three types as Table 1.2.

The MHSR train uses a superconducting magnet to float the vehicle and obtain propulsion power by periodically changing the direction of the magnetic pole. In addition to its high speed, the MHSR train has the characteristics of no noise, no vibration and energy saving. It is expected to become the main means of transportation in the twenty-first century.

1.2.3 The Third Leap: Reducing Air Resistance Brings the Birth of SSR

The third type of HSR: Super-speed Rail (SSR). When the train runs in vacuum, there is no limit to the operating speed. In order to reduce the air resistance, the third leap has been made. SSR, the third type of HSR, is the suspension of HSR in the vacuum pipeline. MHSR trains and SSR trains are as shown in Fig. 1.11.

Third Leap

Fig. 1.11 MHSR and SSR trains

The SSR was produced to satisfy human beings requirements for fast travel, based on the concept of vacuum pipeline by reducing air resistance. The SSR runs in the vacuum pipeline without air and frictional resistance, and the running speed can reach more than 1200 km/h. In fact, there is no air resistance and frictional resistance in vacuum pipeline, so the SSR train can operate "arbitrarily" and speed up to 10,000 km/h. The SSR system is as shown in Fig. 1.12.

SSR is a vacuum pipeline suspended HSR. It is mainly a HSR transportation system suspended in a vacuum pipeline. So SSR also can be called vacuum high-speed rail. The main features are as follows:

① The operation speed of SSR is 1200 km/h (acoustic velocity is 340 m/s).
② SSR has no limit of warning threshold due to no restrictions.
③ SSR has no friction resistance and no air resistance.

The SSR is a vacuum pipeline type of HSR. SSR is a kind of transportation system designed with the principle of "vacuum steel pipe transportation" as the core of the theory. It has the characteristics of ultra-speed, high safety, low energy consumption, low noise and low pollution. The super train may be a new generation of transportation

Fig. 1.12 SSR system

Table 1.3 Types of SSR

Number	Types	Speed/(km/h)	Name	Main countries	Remarks
1.	First	1000–1200	Low-sonic velocity SSR	USA	Sonic velocity: 340 m/s, 1224 km/h
2.	Second	1200–10,000	Normal-sonic velocity SSR	NA	
3.	Third	>10,000	High-sonic velocity SSR	NA	

vehicles after the car, ship, train and aircraft as the fifth type of transportation. This book divides the SSR into the following three types as shown in Table 1.3.

Vacuum piping is an unavoidable choice to reduce air resistance. Ultra speed is the demand for high-speed ground transportation in the twenty-first century. This is determined by two factors. On the one way, from the perspective of environment protection Perspective, SSR is more environmentally friendly than other surface traffic. The carbon dioxide emissions are much lower than those of automobile 100 g/(person·km), aircraft 140 g/(person·km), rail 20 g/(person·km). On the one way, from fast perspective, SSR can achieve the speed of social expectation. And vacuum (or low-pressure) pipeline is the only way to reach an ultra-speed. Therefore, in the future, SSR will be an unavoidable choice.

1.3 Different Types of HSR

In addition to WHSR, HSR also includes MHSR using magnetic levitation technology and SSR operating in a vacuum track. The early warning threshold of HSR of operating speed is mainly based on the energy consumption of HSR trains and the degree of damage to the environment. From the perspective of speed and economy, the warning threshold of WHSR is 400 km/h and the warning threshold of MHSR is 1000 km/h. According to the operating principle and early warning threshold, the book divides HSR into three types:

(1)   The first type of HSR is WHSR. WHSR transportation system is mainly running on the track. It is also called the conventional high-speed rail. The operating speed of WHSR is 200–400 km/h. WHSR has frictional resistance and air resistance. WHSR train is as shown in Fig. 1.13.

(2)   The second type of HSR is MHSR. MHSR is the magnetic suspension type HSR which is mainly suspended on rails to run. MHSR is also called a superconducting high-speed rail. The operating speed of MHSR is 400–1000 km/h. MHSR has air resistance but no frictional resistance. Among them, German MHSR train is as shown in Fig. 1.14.

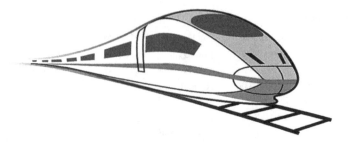

Fig. 1.13 WHSR train

Fig. 1.14 German MHSR train

(3) The third type of HSR is SSR. SSR is a MHSR transportation system suspended in a vacuum pipeline. SSR is also called a vacuum high-speed rail. The operating speed of SSR is more than 1200 km/h (Sonic velocity is 340 m/s) without friction resistance and air resistance. SSR train is as shown in Fig. 1.15.

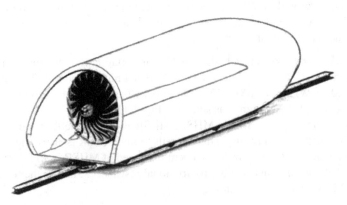

Fig. 1.15 SSR train

1.4 Characteristics of HSR

The rapid development of HSR is determined by its own characteristics, especially its unique technological advantages, which are not available in other modes of transportation (cars, airplanes, traditional trains, etc.). Japan's WHSR train is as shown in Fig. 1.16.

(1) High speed. The test speed of high-speed rail has exceeded 603 km/h and the maximum operating speed is above 350 km/h. Especially, the speed of SSR not only surpasses the sonic velocity (340 m/s), but it will rewrite human history. Compared with other modes of transportation (Fig. 1.17), the advantages of high-speed rail are obvious.

(2) Large volume. HSR is currently mainly used for passenger transportation. At present, no other kind of transportation can surpass HSR in passenger volume. For example, the maximum number of passengers per year on a highway will not exceed 10 million. But according to Japan's statistics, a HSR line has carried 150 million passengers per year. The passenger volume of HSR is as shown in Fig. 1.18.

(3) High safety. In all types of transportation, HSR is the safest. According to statistics, the number of deaths per billion people per kilometer is 1.971 for railways, 18.929 for cars and 16.006 for airplanes. There are about 50,000

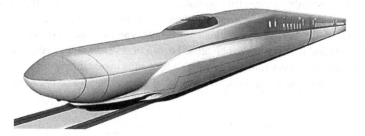

Fig. 1.16 Japan's WHSR train

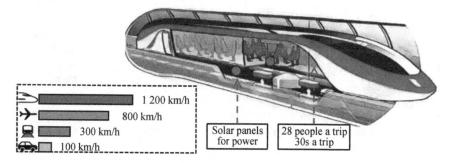

Fig. 1.17 The speed of different vehicles (picture from the network)

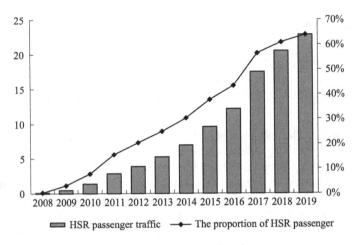

Fig. 1.18 Number of HSR passengers

deaths and 1.7 million injuries per year due to road traffic accidents in Europe, 125 times more than that of traditional railway. And in the United States, there are about 50,000 deaths in highway accidents,while there are less than 100 deaths in railways. In terms of safety, HSR is better than cars and airplanes as shown in Table 1.4.

(4) High punctuality rate. HSR operation is controlled by computer, and only receives signals from the vehicle during operation. Hence, the bad weather like big wind, fog has almost no impact on it unless an earthquake occurs. Airplane airports and highways must be closed down in severe weather conditions such as dense fog, heavy rain and snow. The SSR runs inside the pipeline, so the bad weather such as wind, rain, fog has no influence on it. The high punctuality

Table 1.4 List of global accidents of HSR

Number	Time (Year)	Country	Deaths/Person	Injuries/Person	Reasons
1.	1998.06.03	Germany	101	88	Tyre fracture
2.	2002.11.16	France	12	10	Line short
3.	2005.04.25	Japan	107	562	Overspeed derailment
4.	2011.07.23	China	41	38	Signal failure
5.	2013.07.24	Spain	80	170	Overspeed
6.	2015.11.14	France	11	32	Overspeed
7.	2018.10.21	Taiwan, China	18	207	Overspeed derailment

Table 1.5 The impact degree of different environments on vehicles

Factors		HSR train	Airplane	Car	Traditional train
Natural factors	Wind	Little	Little	Huge	Huge
	Rain	Little	Little	Huge	Huge
	Lightning	Some	Some	No	Some
	Temperature	Little	No	Little	Huge
	Debris flow	Some	No	Some	Some
	Earthquake	Some	No	Some	Some
Human factors	Throwing foreign objects	Some	No	Some	Some

rate (Table 1.5) is also one of the reasons why the HSR is popular among passengers.

(5)   Less pollution. The electrified HSR has no dust, soot and other exhaust gas pollution. Although construction of power plants will cause pollution, the amount of pollution is less than that of freeways and air transportation. The ratio is about 1:3:4. In addition, the noise of HSR is 5–10 dB (decibel) smaller than that of freeways. Environmental pollution degrees of different vehicles are as shown in Table 1.6.

(6)   Land conservation. Compared with the four-lane freeway, the area of HSR is only half of that of the freeway, and most HSR lines are built on dedicated bridges, which will not obstruct the ground transportation. A freeway with 8 lanes can realize the volume of one HSR. While the rail covers an area of 13.8 m wide, the six-lane highway covers an area of 37.5 m wide, as shown in Fig. 1.19.

(7)   Low energy consumption. HSR takes the least energy consumption among all types of transportation. The energy consumption per person per kilometer is only 3.6 kW·h, which is equivalent to 10% of the energy consumption of the aircraft. Energy consumption ratio of various modes of transportation is shown in Fig. 1.20.

Table 1.6 Environmental pollution degrees of different vehicles

Name		Road transportation	Air transportation	HSR transportation
Emission substance	CO	1.26	0.51	0.003
	NO_x	0.25	0.7	0.1
	CO_2	111	158	28
	SO_2	0.03	0.05	0.01
Noise (internal)	dB	76	81	68

Unit: g/(person·km)

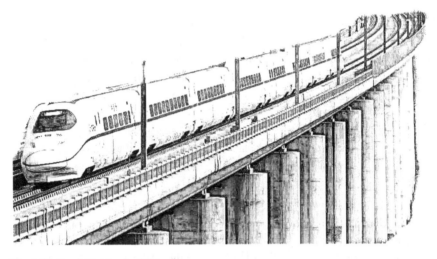

Fig. 1.19 Special bridge for HSR

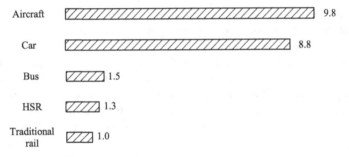

Fig. 1.20 Energy consumption percentage of various modes of transportation

(8) High comfort. The HSR train is luxuriously arranged with complete working and living facilities, spacious and comfortable seats, good running performance, large activity space and stable operation. The train is very quiet because of shock absorption and sound insulation. Travelers are comfortable, convenient and enjoyable when travel on a high-speed track. No matter they are lying or walking, it is more comfortable than stay in other modes of transportation. Business seats of the HSR train are shown in Fig. 1.21. Advantages of HSR are shown in Table 1.7.

HSR has many advantages. In addition to the above advantages, it also has its social, economic, environmental and external boundary benefits.

Fig. 1.21 Business seats of HSR train

Table 1.7 Advantages of HSR

Advantages	Details	
Large transportation capacity	A long train can transport more than 1000 people	Operating interval is 3 min
Strong adaptability to the natural environment	All-weather running	Rarely affected by rain, snow and fog
Short departure interval	Take the "bus" model	Travelers can set off at any time
Energy saving and environmental protection	Green transportation	Energy conservation

1.5 Summary

Since Japan built the world's first High-speed Rail (HSR) in 1964, HSR has developed rapidly. The word will enter the "era of high-speed rail" by 2020. Therefore, "A Brief History of High-Speed Rail" introduces the emerging concept of "high-speed rail" systematically to readers. It includes the terminology, structure function and development trend of high-speed rail system. It is designed to allow readers to fully understand the past, present and future of HSR.

Chapter 2
Conceptual Terminology of HSR

As a safe, reliable, fast, comfortable, large-capacity, low-carbon and environmentally-friendly transportation vehicle, HSR has become the mainstream transportation in the world, leading humanity to a new era. According to the statistics of the International Railway Union, as of December 31, 2021, the total operating mileage of HSR in all countries of the world was 50,000 km, the mileage under construction was 10,000 km, and the planned mileage was 20,000 km. The operation mileage of HSR in some countries is as shown in Table 2.1.

There is no uniform definition of the term "high-speed rail", and different organizations or countries have different definition standards for it. However, in recent years, the standards of various places have been close to each other. The International Railway Union advised to set a standard definition that the high-speed rail refers to the line whose design speed is 200 km/h by transforming the original line, or the new line whose design speed is more than 250 km/h. This book defines "high-speed rail" based on the 3S theory (Speed, Space and Service). HSR should meet the 3S theory: speed is high, space is large and service quality is high. China's WHSR train is as shown in Fig. 2.1.

HSR is a system, called HSR system for short. It includes the narrow definition of HSR and the broad definition of HSR.

① HSR in the narrow sense refers to the traditional WHSR transportation system, which is also the most common understanding known as conventional HSR.
② HSR in the broad sense includes the traditional WHSR transportation system, the MHSR transportation system and the SSR transportation system. The classification of the broad HSR system is shown in Table 2.2.

2.1 Overview of HSR

HSR plays a pivotal role in the development of the economy around the world. This is determined by two reasons: On the one hand, HSR has greatly expanded the capacity

© Southwest Jiaotong University Press 2023
Q. Hu and S. Qu, *A Brief History of High-Speed Rail*,
https://doi.org/10.1007/978-981-19-3635-7_2

Table 2.1 Operation mileage of HSR in some countries

Countries	Chinese mainland	Japan	France	Germany	Spain	Italy	Korea
Operation mileage/km	40,941	3446	2793	3368	4900	1432	1530

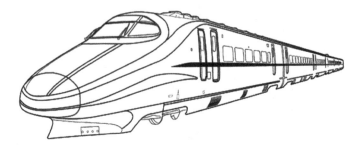

Fig. 2.1 China's WHSR train

Table 2.2 Types of HSR

Number	Types	Name		Speed/(km/h)	Remarks
1.	First	WHSR	Low-speed WHSR	200–300	Wheel High-speed Rail
			Normal-speed WHSR	300–350	
			High-speed WHSR	350–400	
2.	Second	MHSR	Low-temperature MHSR	400–500	Magnetic High-speed Rail
			Normal-temperature MHSR	500–800	
			High-temperature MHSR	800–1000	
3.	Third	SSR	Low-sonic SSR	1000–1200	Vacuum pipe magnetic suspension high-speed rail
			Normal-sonic SSR	1200–10,000	
			Super-sonic SSR	>10,000	

of rails and greatly reduced the time between cities. It has brought great convenience to people when they travel. On the other hand, HSR has greatly improved the transport capacity and further improved the quality of service. However, there are two basic conditions for the formation of HSR. Only meet these two conditions can HSR appear and survive. The first one is a densely populated city with a high standard of living, which can withstand relatively expensive fares and multiple stops. For example, European continents such as Paris, France and Berlin, Germany and the dense cities of Japan are the birthplace of HSR. The second one is a higher social-economic and scientific basis that guarantees the construction, operation and maintenance needs of

HSR, such as Japan, Germany. Through years of development, the HSR has formed its own development modes in Japan, France, Germany and China.

(1)   Japan's Shinkansen is the single line mode. The single line mode is to build all new lines dedicated to passenger trains. Mainly operated by single line but failed to form a network. As shown in Fig. 2.2. On October 1, 1964, the Tokaido

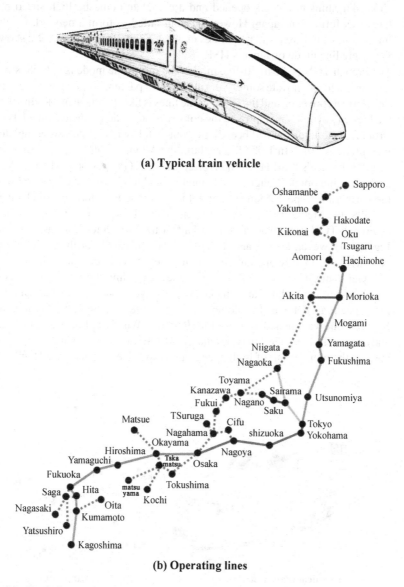

(a) Typical train vehicle

(b) Operating lines

Fig. 2.2 Single line mode of Japan's HSR

Shinkansen was officially opened with operating speed of 210 km/h. It transports 360,000 passengers per day and has annual passengers of 120 million. This electrified standard gauge two-lane rail, dedicated to passenger transport, represented the first-class HSR technology in the world at that time. Between 1975 and 1985, HSR lines such as the Sanyang Shinkansen, the Tohoku Shinkansen, and the Joetsu Shinkansen were opened. In 1997, the Hokuriku Shinkansen was opened and a complete domestic high-speed rail lines was formed in Japan. However, Japan failed to form a network for the limitation of land area, and it operates mainly in a single line. Figure 2.2 shows the single line mode of Japan's HSR.

(2)   The French TGV is the multi-line mode. The multi-line mode is to build some new lines and to renovate some of the old lines for passenger transportation. The lines have a core point and the shape of the lines is like the shape of "+". In 1971, the French government approved the construction of the southeast line of TGV (from Paris to Lyon). It had been built from October 1976, and was completed in September 1983. In 1989, France built the Atlantic high-speed rail line and in 1993, France's third HSR, the Nordic line of TGV was opened from Paris as the starting point through the Channel Tunnel to London. It connected with the northern European countries, and it is an important international channel. In 1999, the Mediterranean line was completed. From then the existing lines of French TGV can cover more than half of the French territory as shown in Fig. 2.3. However, France also failed to form a network for the limitation of land area. It is mainly operated in a multi-line format in the shape of " + ".

(3)   The Germanic ICE is the mixed-line mode. Regarding the mixed line mode, all routes are newly built and shared by passengers and freight. The German high-speed rail ICE was first commissioned in 1985. In 1991, the Mannheim-Stuttgart line was opened. In 1992, the Hannover-Wurzburg line was completed and opened. In 1992, Germany purchased 60 ICE trains, 41 of which were operated on the number 6th of high-speed rail, respectively connecting Hamburg,

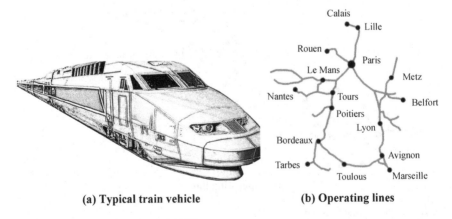

(a) Typical train vehicle   **(b) Operating lines**

Fig. 2.3 Multi-line mode of French TGV

Frankfurt and Stuttgart. At present, Germany's pan-European HSR and the third phase of HSR have been built one after another, achieving the international direct transportation of HSR as shown in Fig. 2.4.

(4) China Railways High-speed (CRH) is a network model. The network mode is to establish some new lines and partially transform the old lines for passenger transportation. It has formed a HSR network for convenient transfer. On August 1, 2008, the first HSR was operated from Beijing to Tianjin. Since then, China's HSR has developed rapidly. By the end of December 2015, CRH has the longest operating mileage and the largest scale under construction. China is the only country that has the network of "four vertical lines and four horizontal lines". By December 2020, China's HSR will have a mileage of more than 30,000 km and China will form a HSR network with the "eight vertical lines and eight horizontal lines". CRH adheres to the combination of original innovation, integrated innovation, introduction, digestion, absorption and re-innovation. It has built a HSR technology system with independent intellectual property rights. CRH has reached the world's advanced level, as shown in Fig. 2.5.

Fig. 2.4 Mixed line mode of Germanic ICE

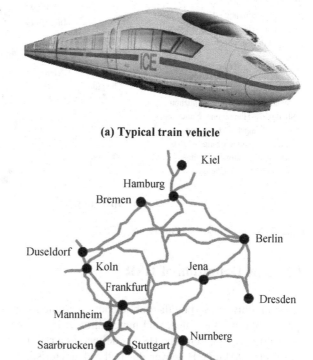

(a) Typical train vehicle

(b) Operating lines

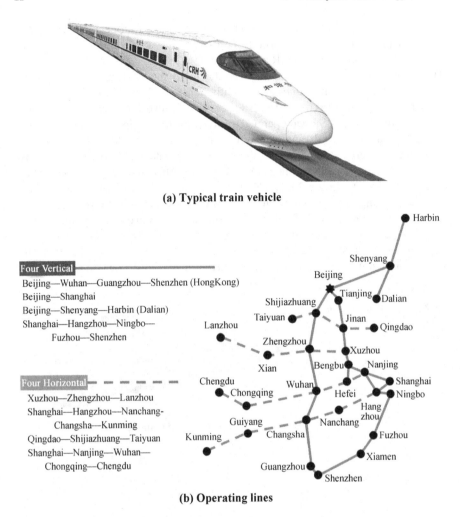

(a) Typical train vehicle

Four Vertical

Beijing—Wuhan—Guangzhou—Shenzhen (HongKong)
Beijing—Shanghai
Beijing—Shenyang—Harbin (Dalian)
Shanghai—Hangzhou—Ningbo—
    Fuzhou—Shenzhen

Four Horizontal

Xuzhou—Zhengzhou—Lanzhou
Shanghai—Hangzhou—Nanchang-
    Changsha—Kunming
Qingdao—Shijiazhuang—Taiyuan
Shanghai—Nanjing—Wuhan—
    Chongqing—Chengdu

(b) Operating lines

Fig. 2.5 Network model of China's HSR (CRH)

2.2 The Definition of HSR

HSR is highly praised by the public and governments around the world for its high speed, convenience, safety and punctuality. It is also an important part of current transportation research. The HSR is defined as a rail line with an operating speed of more than 200 km/h. However, due to the characteristics of HSR itself, the definition of it in the world is not unified. According to the HSR technology used over the world, the definition can mainly be divided into three criteria: the European Union, the UN Economic Commission and the International Railway Federation. At the same time, different countries have different definitions of HSR. China adopts the

Table 2.3 Definition of HSR

Name		WHSR		HSR trains
First	International railway federation	New lines	Design speed 250 km/h or more	Trains with commercial operation above 250 km/h
		Transform old lines	Design speed 250 km/h or more	Trains with commercial operation above 200 km/h and high quality of service
Second	European Union	New lines	Allowable speed 200 km/h or more	Operation speed above 250 km/h and can reach 300 km/h
		Transform the old lines	Allowable speed 200 km/h or more	Trains with operation speed above 200 km/h
Third	UN economic commission	New lines	Allowable speed 200 km/h or more	Trains with design speed above 250 km/h
		Transform the old lines	Allowable speed 200 km/h or more	Trains with design speed above 250 km/h
Fourth	Japanese standard	New lines	Allowable speed 200 km/h or more	Trains with design speed above 250 km/h
		Transform the old lines	Allowable speed 200 km/h or more	Trains with design speed above 250 km/h
Fifth	Chinese standard	New lines	Line speed 200–250 km/h	Trains with operating speed above 250 km/h and high-speed rail trains with operating speed above 300 km/h
		Transform the old lines	Line speed 20–250 km/h	

definition of HSR made by International Railway Union. The various standards are shown in Table 2.3.

2.2.1 The Definition of WHSR

The International Railway Federation is a non-government rail joint organization involving some railway institutions and related organizations in some European countries and other continents. Later, it recruited some organization of non-European countries. Its purpose is to promote the development of international railway transportation, promote international cooperation and improve railway technical equipment and operation methods. It carries out scientific research on related issues and realizes the unification of technical standards for railway buildings and equipment. The definition of WHSR in this book is made by Chinese standards.

(1) International Railway Federation Standards. According to the definition of the International Railway Federation, the WHSR refers to the rail system with an operation speed above 200 km/h by transforming the old line (linearization, gauge standardization) or the rail system with new lines' operating speed is above 250 km/h. In addition to the speed standards for train operation, the vehicles, rails and operations also need to be upgraded.

① HSR requires that the design speed of the new lines is above 250 km/h or the transformed old lines (linearization, gauge standardization) have a design speed above 200 km/h or even reach 220 km/h.

② HSR motor train is the train which has a commercial operating speed above 250 km/h or has a commercial operating speed of 200 km/h but has high service quality such as tilting trains.

(2) European Union standards. In the process of establishing the Trans-European High-speed Rail Network (TENR) system, the European Union proposed the definitions of "high-speed rail" "high-speed rail motor trains" and issued the Directive "96/48/EC". The Directive (D + IRECTIVE 96/48/EC) gives the standards for both "high-speed rail" and "high-speed rail motor trains". This standard is now generally applicable to European Union member states.

① HSR has a design speed above 250 km/h for the new lines or has a design speed above 200 km/h for transformed old lines.

② HSR motor train has the running speed above 250 km/h and sometimes the speed can reach up to 300 km/h on the new lines and on the existing rails, the running speed can reach 200 km/h.

(3) Standards of the UN Economic Commission. According to the United Nations Economic Commission for Europe (UNECE) Transport Statistics Working Group, the standards of "high-speed rail" and "high-speed rail trains" have been set like the European Union. In 1985, "International Railway Route Agreement" signed by the United Nations Economic Commission for Europe in Geneva stipulated that the new lines for mixed passenger transportation and freight transportation (refer to as passenger and cargo line) should have a speed above 250 km/h, and the new lines for passenger transportation (referred to as passenger lines) should have a speed above 350 km/h.

① HSR has a design speed above 250 km/h for the main lines or has a design speed above 200 km/h for transformed old lines.

② HSR motor train has a design speed above 250 km/h or a design speed above 200 km/h for transformed old lines or the fastest speed of 200 km/h of traditional HSR train.

(4) Japanese standards. Japan was the first country in the world to develop the HSR. The Japanese government issued Decree No.71 in 1970, which defined HSR as: where the main section of a railway has a maximum speed of 200 km/h or more, it can be called the high-speed rail.

    ① HSR has an allowed speed for the main lines above 250 km/h or has an allowed speed above 200 km/h for transformed old lines.

    ② HSR motor train has a design speed above 250 km/h on special lines or a design speed of titling trains above 200 km/h for transformed old lines or the fastest of speed 200 km/h for traditional HSR motor train.

(5) Chinese standards. China has become a country owing the most complete technology, the strongest integration capability, the longest operating mileage, the highest operating speed and the largest scale of construction around the world. The "High-speed Rail Design Code (TB10621-2014)", which was piloted in 2009, stipulates that HSR refers to a new rail with a maximum speed above 250 km/h for passenger transportation. The Regulations on Railway Safety Management (Supplementary Provisions) which was implemented on January 1, 2014 stipulates that HSR refers to rails with an operating speed above 250 km/h (including reservations) and an initial operating speed above 200 km/h for passenger transportation (refer to as passenger lines).

    ① HSR has a design speed for the new lines above 250 km/h or has a design speed above 200 km/h for transformed old lines. The trains operating on the lines with a speed below 250 km/h are called "trains". The trains with a speed from 300 to 350 km/h on the lines are called "high-speed rail trains". China's HSR construction has advanced by leaps and bounds, and the scale of the HSR network has expanded rapidly (Fig. 2.6).

Fig. 2.6 China's HSR network topological graph

Fig. 2.7 CRH train

    ②   The types of China's HSR motor trains are as follows: CRH1, CRH2, CRH3, CRH5, CRH6, CRH380A, CRH380B, CRH380C, CRH380D, CR200, CR300, CR400, etc. (Fig. 2.7).

2.2.2 The Definition of MHSR

MHSR train is a modern high-technique vehicle that uses the electromagnetic force to realize the contactless suspension between the train and the track and guidance. And then it uses the electromagnetic force generated by the linear motor to pull the train. Therefore, this book refers to the magnetic suspension rail as the MHSR. MHSR trains and motors work exactly in the same way. The "rotor" of the motor is placed on the train and the "stator" of the motor is laid on the track. Through the interaction between the "rotor" and the "stator", electrical energy is converted into forward kinetic energy. When the "stator" of the motor is energized, the "rotor" can be rotated by the action of the current on the magnetic field. When the power is transmitted to the "stator" of the orbit, the train acts like a "rotor" of the electric motor to do a linear motion by the effect of the current on the magnetic field. Figure 2.8 shows the Japan's MHSR train.

    The MHSR train is mainly composed of three parts: the suspension system, the propulsion system and the guiding system. Although it is possible to use a propulsion system that is independent of the magnetic force, in most of the current designs, the functions of these three parts are all done by magnetic force.

(1)   The suspension mode of MHSR train. When the magnet passes over a piece of metal, the electrons on the metal begin to move as the magnetic field changes. The electrons form a loop and then produce their own magnetic field. Because the same electric charge mutual repulsion, and opposite electric charge attract, moving the magnet over metal results in a push-up force on the moving magnet. If the magnet moves fast enough, this force will be large enough to overcome the gravity and lift the moving magnet. Among them, the superconducting MHSR

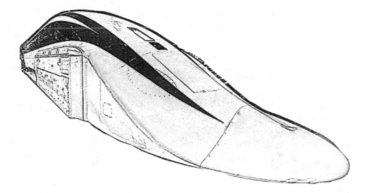

Fig. 2.8 Japan's MHSR train

uses the strong repulsive force between the electromagnetic field formed by the superconducting electromagnets and the electromagnetic field formed by the coil on the rail to levitate the vehicle (Fig. 2.9).

(2) The guiding method of MHSR train. MHSR train uses the action of electromagnetic force to guide. There are a constant-conduction magnetic guiding system and a superconducting magnetic repulsive guiding system.

    ① The constant-conduction magnetic guiding system is similar to the suspension system. It installs a set of electromagnets on the side of the trains to guide. There is a certain gap between the train body and both sides of the guide rail. When the train is offset to the left or right, the guiding electromagnet on the train interacts with both sides of the guide rail to return the train to its correct position. The control system maintains this lateral clearance by controlling the current in the guiding magnets to achieve the purpose of controlling the direction of train's travel (Fig. 2.10).

    ② The guiding system of superconducting magnetic repulsion guiding system can be constituted in three ways. (a) The first mode is to install

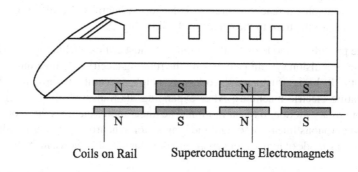

Fig. 2.9 Levitation mode of MHSR

Propulsion Coill   Levitation/   Side Wall   Wheel Bearing
                   Guidance Coil   Beam   Surface

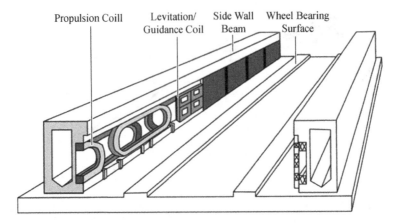

Fig. 2.10 Guiding method of normal conduction magnetic train

a mechanical guide on the train to achieve guidance. This device typi-
cally employ a side-guided auxiliary wheel on the train that interacts with
the side of the guide rail (rolling friction) to create a restoring force that
balances the lateral forces generated by the train as it travels along the
curve, thereby making the train to run along the guide rail. (b) The second
mode is installing a special guiding superconducting magnet on the train
to generate a magnetic repulsion force with the ground coil and the metal
belt on the side of the guide rail, and the force is balanced with the lateral
force of the train to keep the train in the correct state. This guiding method
avoids mechanical friction and allows the train to maintain a certain lateral
clearance as long as it controls the current in the lateral ground-guiding
coil. (c) The third mode is guided by magnetic force. The "zero magnetic
flux" guide is a closed coil with a shape of "8". When the superconducting
magnet set on the train is located on the symmetrical center line of the
coil, the magnetic field in the coil is zero. When the train has a lateral
displacement, the magnetic field in the "8" shaped coil is zero and then
generates a reaction force to balance the lateral force of the train and
return the train to the center of the line (Fig. 2.11).

(3)   The propulsion mode of MHSR train. The most critical technology of MHSR
train propulsion system is to expand the rotating electric machine into a linear
motor. Its basic structure and working principle are similar to those of ordinary
rotating electric machines. After expanding, its transmission mode changes
from rotary motion to linear motion. The linear motor includes the short stator
asynchronous linear motor and the long stator synchronous linear motor. The
working mode of long stator synchronous linear motor is shown in the Fig. 2.12.

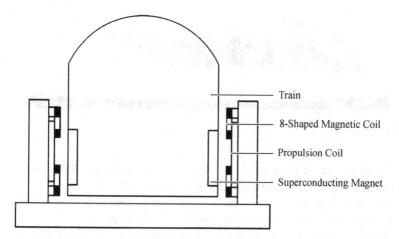

Train

8-Shaped Magnetic Coil

Propulsion Coil

Superconducting Magnet

Fig. 2.11 Guiding method of superconducting magnetic train

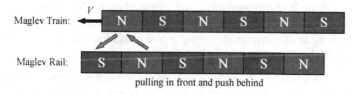

Maglev Train:

Maglev Rail:

pulling in front and push behind

Fig. 2.12 Working mode of long stator synchronous linear motor

2.2.3 The Definition of SSR

Super-speed Rail (SSR) is a kind of transportation designed with the core principle of "vacuum pipeline transportation". It is also called Hyperloop or Pneumatic Tubes. SSR has the characteristics with ultra-speed, high safety, low energy consumption, low noise and low pollution. SSR trains may be a new generation of transportation vehicles after cars, ships, trains and airplanes. In 2013, Elon Musk proposed the "Hyperloop" program. He believes that the SSR can carriage passengers at an ultra-speed of 1200 km/h. Therefore, SSR is regarded as the development direction of future traffic, attracting more and more countries to research and develop. Figure 2.7 shows American SSR–Hyperloop.

The SSR system is to build a pipeline that is isolated from the outside air. After pumping the pipeline into a vacuum, the MHSR train is operated in it (Based on the idea of Musk, the schematic diagram of the SSR is shown in Fig. 2.13). In an environment with little friction, the trains in the low-pressure pipeline can operate at a speed of 1200 km/h. From the characteristics of the existing five modes of transportation (rail, aviation, water transport, roads, pipelines, etc.), the SSR has some characteristics of five kinds of vehicles:

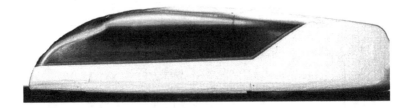

Fig. 2.13 American SSR—Hyperloop

① The first mode of transportation—pipeline transportation. SSR is traveling in pipelines and has the characteristics of pipeline traffic.

② The second mode of transportation—rail transportation. SSR uses magnetic levitation technology and has the features of rail transportation.

③ The third mode of transportation—road transportation. The transportation capacity of SSR is equivalent to the transport capacity of bus (20 to 50 passengers on the public bus) so SSR has the characteristics of road transportation.

④ The fourth mode of transportation—air transportation. SSR operates at a speed similar to that of an airplane so it has the characteristics of air transportation.

⑤ The fifth mode of transportation—water transportation. The SSR train floats in the air so it has the characteristics of water transportation.

Therefore, SSR is a brand new type of transportation that combines the characteristics of five existing modes of transportation. It may also be the sixth type of transportation. Figure 2.14 shows the structure of SSR.

(1) SSR train. SSR is a new type of transportation based on the concept of "vacuum pipe transportation". This kind of vehicle is a new generation after cars, ships, trains and airplanes. It has the characteristics of ultra-speed, high safety, low energy consumption, no noise and zero pollution. Due to the use of vacuum pipeline and magnetic levitation technology, this book suggests to call the vehicle as vacuum train or super-speed train. The simplified diagram of super-speed trains is shown in Fig. 2.15.

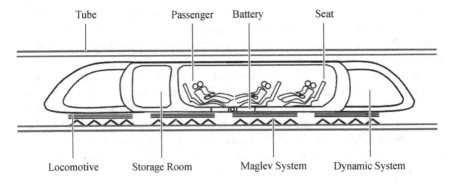

Fig. 2.14 The structure of SSR

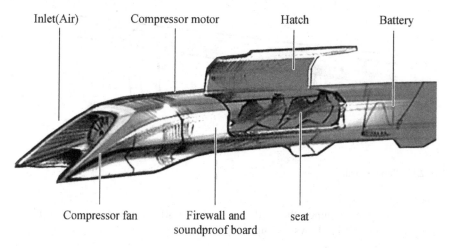

Inlet(Air) Compressor motor Hatch Battery

Compressor fan Firewall and soundproof board seat

Fig. 2.15 Simplified diagram of SSR train

SSR is also called the vacuum pipeline magnetic train (refer to as the vacuum maglev train). It is a train that has not yet been built and may be the fastest transportation in the world. This kind of trains runs in a vacuum pipeline which is not affected by air resistance, friction and weather. The cost of SSR is lower than that of the traditional rail. The speed can reach 4000–20,000 km/h, which is several times fasten than the speed of aircraft, while the energy consumption is many times lower than that of the aircraft. In the future, SSR may become the fastest means of transportation in the twenty-first century. Figure 2.16 shows the simulation diagram of SSR train.

(2) Vacuum piping. Unlike the traditional rail, the SSR is a new vacuum suspension frictionless flight system. The super-speed rail system is consisted of transport pipelines, manned cabins, vacuum equipment, suspension components, ejection and braking systems as shown in Fig. 2.17.

① The operating characteristics of SSR: running in the pipeline suspended with no resistance and the speed can reach above 1000 km/h.

② Transmission method of SSR: The SSR train floats in the vacuum-treated pipeline because of the magnetic technology and then uses the catapult

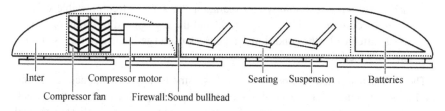

Inter Compressor motor Seating Suspension Batteries

Compressor fan Firewall:Sound bullhead

Fig. 2.16 Simulation diagram of SSR train

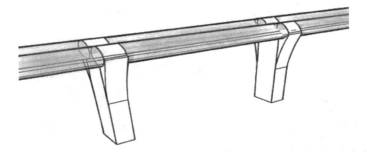

Fig. 2.17 Super pipeline

device to launch the SSR train to the destination without interruption. Figure 2.17 shows the super pipeline.

2.3 Speed Definition of HSR

The operating speed of HSR has practical significance for passengers. Operating speed of HSR includes multiple concepts such as maximum operating speed, average travel speed and the speed of trains in the tunnel, and each concept has practical significance. The highest test speed should be at least 10% higher than the real maximum operating speed to ensure safety. Because of the different operation modes of HSR, the experimental speeds of HSR are also different.

(1) The highest test speed. For any transportation systems, the highest test speed is designed based on special planning and external conditions such as lines, power boosts, special signals and vehicle equipment. The speed is usually achieved under special operation methods and safety precautions.

(2) The highest operating speed. The highest operating speed is designed and operated under everyday conditions. This speed is such that the entire system of HSR–the structure, the vehicle, the control, the support, etc., can operate under everyday conditions and withstand passenger rides and weather changes, and must be handled by specialized personnel.

(3) The highest design speed. This speed is also known as the calculated driving speed. It refers to the maximum driving speed at which the driver of the medium driving technology can maintain the safe and comfortable driving when the climatic conditions are good and the high-speed rail train operation is only affected by the conditions of the track itself (geometric elements, tracks, ancillary facilities, etc.).

The highest test speed is important in the assessment of system characteristics and development potential. However, the maximum operating speed defines the actual, achievable performance of the system. The difference between the highest test speed and maximum operating speed is very large. The highest test speed may be 50%

Table 2.4 Test speed of HSR in some countries

Number	Countries	Time (Date)	Vehicles	Test speed/(km/h)	Remarks
1.	France	03-04-2007	TGV-V150	574.8	WHSR
2.	China	09-01-2011	CRH380BL	487.3	
	China	03-12-2010	CRH380AL	486.1	
	China	28-09-2010	CRH380A	416.6	
3.	Japan	26-07-1996	Type 955	443	
	Japan	04-12-1993	Type 952,Type 953	425	
4.	Germany	01-06-1988	ICE	406	
5.	Japan	21-04-2015	JR Maglev L0 Series	603	MHSR
	Japan	02-12-2003	JR Maglev MLX01	581	
6.	Germany	12-11-2003	TransrapidTR-08	501	

Table 2.5 Maximum operating speed of HSR in some countries

Number	Countries	Time (date)	Vehicles	Maximum operating speed/(km/h)	Remarks
1.	France	03-01-2007	TGV	320	WHSR
2.	China	01-08-2008	CRH	350	
3.	Japan	05-07-1967	Type 955	320	
4.	Germany	03-10-1999	ICE	300	

to 80% higher than the maximum operating speed. The test speed and maximum operating speed of some HSR trains in some countries are shown in Tables 2.4 and 2.5.

The maximum operating speed is the most important measure index of HSR. From the 1960s to the 1970s, the highest test speed increased from 250 km/h to 350 km/h. Germany achieved a major breakthrough in 1988. ICE train's test speed reached up to 406 km/h. Then, France achieved a leap in 1991. TGV train created a record speed of up to 515 km/h in the test. At present, among the hundreds of trains operating in many countries every day, the maximum operating speed is 250–350 km/h. The French TGV has recently created a record of 1000 km operating mileage at an average speed of 317 km/h.

Speed is the most comprehensive and critical indicator of HSR system. It is the main indicator to measure the HSR technical level of a country, because speed often refers to the operating speed under a series of indicators such as safety, reliability, economy, energy conservation and environmental protection. From these aspects, operating speed indicates whether HSR technology is at the world's leading level. The speed types of HSR are shown in Table 2.6.

Table 2.6 Speed types of HSR

Number	Types	Name		Speed/(km/h)
1.	First	Test speed	SSR	200–600
			MHSR	400–100
			SSR	1000–10,000
2.	Second	Design speed	SSR	200–500
			MHSR	500–800
			SSR	800–1200
3.	Third	Operating speed	SSR	200–400
			MHSR	400–800
			SSR	>1000

2.4 Summary

In 1964, the world's first HSR was opened in Japan and the first round of "high-speed rail heat" was launched worldwide. However, due to technical problems, HSR was not vigorously developed and the operating speed was lower than 300 km/h. In 1995, after the French HSR technology became the technical standard for all-European high-speed trains, the second round of "high-speed rail heat" was launched worldwide. Due to the economic downturn at that time, especially limited economic capacity of developing countries, HSR only took place among the developed countries. In 2008, due to China's HSR development, the world's third round of "high-speed rail heat" was launched. Since 2015, with the advanced, mature, economic, applicable and reliable HSR technology, HSR has been built and operated in developing countries. Then the fourth round of "high-speed rail heat" has been set up worldwide. Therefore, with the rapid development of HSR, the world will enter the "era of high-speed rail" in 2020 and the world will become a "global village" under the HSR.

Chapter 3
Attribute Characteristics of HSR

At present, HSR is the transportation mode with the highest running speed, the longest operating mileage and the largest carrying capacity. Compared with other modes of transportation (cars, airplanes, traditional trains, etc.), HSR has obvious advantages. Compared with the traditional rail (it is also called ordinary rail in this book), the biggest advantages of HSR are land conservation, high operating speed, low energy consumption, large transportation capacity, excellent industrial structure and good social benefits (Fig. 3.1).

3.1 Transport Capacity

Large transport capacity is one of the main technical advantages of HSR. Statistics show that the maximum passengers of a freeway cannot exceed 10 million a year, while HSR passenger volume reached 150 million per year. For example, the Tokaido Shinkansen in Japan operated 11 trains per hour. There are 283 trains operating one day. Each train can carry 1200–1300 passengers and the average annual passengers are 120 million.

(1) The comparison of the transport capacity of HSR and freeway and aviation. Statistics show that the HSR transport capacity is 10 times of that of aviation and 5 times of that of freeway. However, the HSR transportation cost is only 1/5 of that of the aviation and 2/5 of that of freeway transportation as shown in Table 3.1. For example, in terms of hourly passenger volume, the maximum capacity of two-way HSR can reach 64,000–72,000 passengers, the capacity of four-lane freeways is about 9800 persons and the capacity of two runways airport is about 12,000 persons. It is shown in Fig. 3.2 and Table 3.1.

① HSR transport capacity. At present, HSR in all countries can reach the minimum headway time of 4 min and the Japan can reduce to 3 min. Except for 4 h of maintenance time, the rest time can operate 280 pairs

© Southwest Jiaotong University Press 2023
Q. Hu and S. Qu, *A Brief History of High-Speed Rail*,
https://doi.org/10.1007/978-981-19-3635-7_3

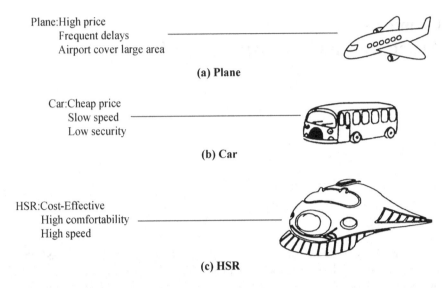

Plane:High price
  Frequent delays
  Airport cover large area

(a) Plane

Car:Cheap price
  Slow speed
  Low security

(b) Car

HSR:Cost-Effective
  High comfortability
  High speed

(c) HSR

Fig. 3.1 Advantages of HSR

Table 3.1 Comparison of transport capacity of different transportation (regard HSR as 1)

Transportation modes	HSR	Freeway	Aviation
Transport capacity (based on HSR)	1	0.2	0.1
Transport cost (based on HSR)	1	2.5	5

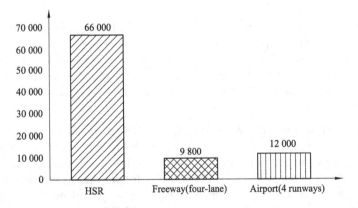

Fig. 3.2 Maximum transport capacity per hour of HSR, freeway and aviation

of trains per day. If each train can carry 800 passengers, the transport capacity of one line can reach 82 million passengers.

②   Road transport capacity. Statistics show that the four-lane freeway can pass through 1250 cars per hour in one-direction. During 20 h throughout the day, it can pass 25,000 vehicles. If large cars account for 20% and each car carry 40 passengers. Small cars account for 80% and each car carry 2 people. The average annual transport capacity can reach 87.6 million in one-direction.

③ Aviation capacity. Statistics show that aviation transportation is mainly limited by airport capacity. If the annual takeoff and landing capacity of a dedicated runway is 120,000, the one-way transportation capacity of a large passenger aircraft can only reach 15–18 million. Daily one-way transport capacity of HSR, freeway and aviation is as shown in Fig. 3.3

It can be seen from Figs. 3.2 and 3.3 that the transport capacity of HSR and freeway and aviation is completely different. The transport capacity of HSR is higher than that of freeway and aviation.

(2) Comparison of the transport capacity of HSR and traditional rail. The minimum design tracking interval of China's HSR is 3 min. The annual one-way transport capacity is 80 million person-times, which means the annual transport capacity can reach 160 million person-times in total. The daily maintenance time of the HSR is 4 h but it actually takes more than 6 h per day including maneuvering and road test. The traditional rail train generally has 18 carriages. The seats for each carriage are generally 108, after subtracting one dining carriage, the passengers of per train is 1786. The traditional train can be over-crowded and the carriage can be added. In addition, there are many stops and the seat utilization rate is high, so the capacity of each carriage is generally 2.5 times larger than that of HSR train! The annual transport capacity of Japan's Tokaido Shinkansen, which has the largest passengers in the world, is 140 million passengers. The annual transport capacity of Beijing—Guangzhou and Beijing—Shanghai HSR is 95 million and 85 million. Annual transport capacity of HSR and traditional rail is as shown in Fig. 3.4.

As seen from Fig. 3.3, the annual transport capacity of HSR is less than that of traditional trains. In fact, the transport capacity of HSR is not smaller than that

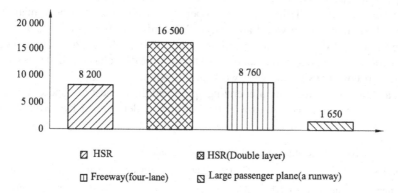

Fig. 3.3 Daily one-way transport capacity of HSR, freeway and aviation

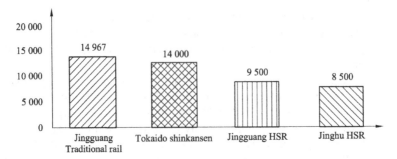

Fig. 3.4 Annual transport capacity of HSR and traditional rail (per ten-thousand passengers)

of traditional rail. With the development of the economy and the improvement of people's living standards, the potential passengers of HSR are very large. In order to further improve the transport capacity of HSR, it is necessary to shorten the departure interval, reduce maintenance time and improve the utilization of seats and so on.

3.2 Speed

Speed is the most important indicator of HSR technology. Countries are constantly improving the speed of trains. The highest operating speed of HSR trains in France, Japan, Germany, Spain, Italy and China has reached 300 km/h, 300 km/h, 280 km/h, 270 km/h, 250 km/h and 350 km/h respectively. If further improvement is made, the running speed can reach 350–400 km/h. Table 3.2 shows the maximum operating speed of HSR in some countries.

The speed of HSR is higher than many other vehicles (cars, ships, traditional rails, etc.), only slower than MHSR and airplane. However, MHSR has the disadvantages in cost and operation, and the airplane is susceptible to weather conditions. Speed comparison between HSR and other vehicles is shown in Fig. 3.5.

At present, China's HSR "350 km/h operation" technology has been matured. All the HSR lines in the world are designed according to the speed of 350 km/h while China's HSR is designed according to the speed of 380 km/h. The test speed of WHSR has reached 575 km/h and the test speed of MHSR has reached 603 km/h. The test speeds of HSR around the world are shown in Table 3.3.

In addition to the maximum operating speed and the highest test speed, passengers are more concerned with travel time, which is determined by travel speed.

Table 3.2 Maximum operating speed of HSR in some countries

Countries	France	Japan	Germany	Spain	Italy	China
Maximum operating speed/(km/h)	300	300	280	270	250	350

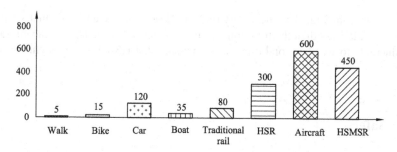

Fig. 3.5 Speeds of different vehicles (km/h)

Table 3.3 The highest test speeds of HSR in some countries

Name	Countries		Highest test speed/(km/h)	Technology
Europe	Germany		406.9	Germanic ICE
	France		574.8	French TGV
	Italy		319	Italian ETR
	Spain		300	French TGV, Spanish TALGO, Germanic ICE
	Britain		300	France TGV
America	American		300	France TGV
Asia	Japan		581	Japan Shinkansen
	Korea		352.4	French TGV-A
	China	Mainland	486.1	TGV, ICE, CRH
		Taiwan	315	Japan's Shinkansen
	Turkey		250	French TGV

Taking Beijing to Shanghai as an example, the travel time for different modes of transportation in normal weather conditions is shown in Fig. 3.6.

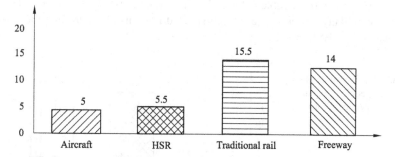

Fig. 3.6 Travel time of different vehicles from Beijing to Shanghai (h)

As seen from Fig. 3.5, the travel time of HSR is comparable to that of the airplane but far less than the traditional rails and freeways. Considering the aircraft is vulnerable to weather conditions, it can be said that HSR is the first choice for travel.

3.3   Safety

Since the advent of HSR, Japan, Germany, France, China and other countries have transported 10 billion passengers totally. Except for several accidents as shown in Table 1.4, there were no major traffic accidents on HSR and no casualties caused by accidents. This is rare in other modes of transportation. In particular, Japan, Germany, France, and China operate thousands of HSR trains every day. Even if five accidents are counted, the accident rate and casualty rate are far lower than other modes of transportation. Therefore, HSR is considered to be the safest means of transportation.

Statistics show that the world's road traffic accidents generally cause 300,000–500,000 deaths per year and the average number of deaths per billion person-kilometer is up to 140. Every year, about 50 crashes of the world's aviation traffic take place causing more than 2000 deaths. The number of deaths per 1 billion person-kilometer is 1.971 of rail, 18.929 of cars and 16.006 of airplane as shown in Fig. 3.7.

The numbers of traffic accidents per 100 million person-kilometers in China are shown according to the data published by the China Academy of Railway Sciences. There are 10.5 people died (severe injuries are 24.88) on the road and 0.29 (severe injuries are 0.72) people died in the railway, which is shown in Fig. 3.8.

Since the biggest danger of WHSR comes from the derailment of the train and the index to evaluate the derailment stability of the wheel is the "derailment coefficient". It denoted by f(Derailment coefficient), that is $f(DC)$. The larger the derailment coefficient, the easier it is to derail. The International Railway Federation requires $f(DC) \leq 1.2$. China's WHSR derailment coefficient safety standard is $f(DC) \leq 0.8$. Table 3.4 shows the derailment coefficient at different speeds in China.

It can be known from Table 3.4 that the current domestically operated WHSR is extremely unlikely to derail due to high-speed driving itself. Therefore, the safety advantages of WHSR are obvious.

Fig. 3.7 Number of deaths per billion people in Japan, 1985 (deaths/one billion)

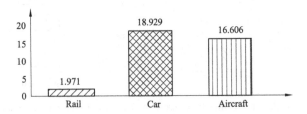

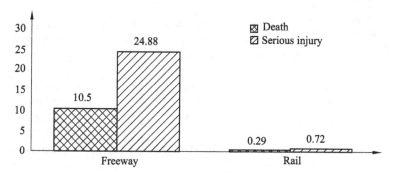

Fig. 3.8 Number of deaths and injuries in traffic accidents per 100 million people per kilometer of roads and rails in China (hundred million deaths/km)

Table 3.4 Derailment coefficient at different speeds in China

Number	Type	Speed/(km/h)	Maximum derailment coefficient	Remarks
1.	CRH2A	250	0.72	$f(DC) \leq 0.8$
2.	CRH2C	300	0.30	
3.	CRH380A	386	0.34	
4.	CRH380A	300	0.13	
5.	CRH380A	350	0.34	

3.4 Punctuality Rate

HSR is controlled by computer. Unfavorable weather such as wind, rain, snow and fog has little effect on the HSR as shown in Table 3.5. HSR trains operate according to the prescribed time so the regularity is stronger than airplanes, cars and other transportation vehicles. In particular, HSR is controlled automatically and can be operated around the clock unless an earthquake occurs. According to the specifications of the Japan's Shinkansen wind speed limit, if a windshield is installed, even in high wind conditions, the HSR train will only slow down. For example, if the wind speed reaches 25–30 m/s, and the train speed limit is 160 km/h. If the wind speed reaches 30–35 m/s (similar to the 11th, 12th level), the train will operate at a speed limit of 70 km/h without stopping as shown in Table 3.6. Aircraft airports and

Table 3.5 Impact degree of weather on the operation of various modes of transportation

Name	HSR	Airplane	Freeway
Wind speed	General	Big	Large
Fog	General	Huge	Huge
Rain	Big	Huge	Huge
Ice	Big	Huge	Huge

	Number	Wind speed (m/s)	Wind speed level	Emergency measures
Table 3.6 Emergency measures for HSR trains in high winds	1.	0–25	1–9	Normal operation
	2.	25–30	9–10	Speed limit: 160 km/h
	3.	30–35	11–12	Speed limit: 70 km/h
	4.	> 35	> 12	Stop operating

freeways must be shut down in severe weather conditions such as fog, heavy rain and ice.

The high punctuality rate is also one of the reasons why HSR is popular among passengers. Reliability of HSR system equipment and high level of transport organization ensure high punctuality to passenger trains. HSR usually sends a train every 4 min. In Japan, the train is sent every 3.5 min at the peak time. Passengers can basically leave at any time without waiting. For the convenience of passengers, the HSR trains are regularly operated and the stations are fixed by trains. This is unmatched by any other kind of transportation. Spain stipulates that the HSR train will return the passenger's full ticket fee if the delay is more than 5 min. Japan stipulates that if the sent time is delayed more than 1 min then the train is regard as delayed and it will return passengers accelerated fee if the delay time is more than 2 h. Therefore, the high punctuality of HSR has won the trust of passengers.

3.5   Energy Consumption

In every country, transportation is a major energy consumer. The energy consumption standards are also important technical indicators for evaluating the advantages and disadvantages of transportation. HSR trains use electric energy and do not consume valuable liquid fuels such as petroleum. According to calculation, the average energy consumption per person-kilometer of various transportation is different as shown in Table 3.7 and Fig. 3.9.

Table 3.7 Energy consumption per person per kilometer for different transportation (g/person/km)

Modes	Traditional train	HSR	Bus	Cars	Airplane
Energy consumption	403.2	571.2	583.8	3309.6	2998.8
Energy	Electric energy	Electric energy	Gasoline or Kerosene	Gasoline or Kerosene	Gasoline or Kerosene

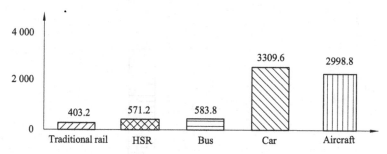

Fig. 3.9 Energy consumption for different transportation (person/km)

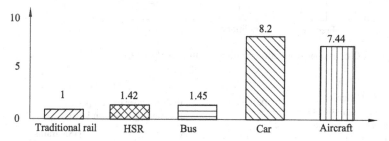

Fig. 3.10 Converted energy consumption index of different transportation

If the energy consumption is compared by the unit of "person/km" and the energy consumption per kilometer of the traditional rail is 1, the converted energy consumption index of different transportation is shown in Fig. 3.10.

3.6 Environmental Impact

The current world requirements for a new generation of vehicles are small impacts on the environment. The energy used by HSR is secondary energy-electricity. The emission generated by HSR is zero and hardly brings environmental pollution. Cars and airplanes use non-renewable energy- gasoline and kerosene that not only pollute the environment, but also produce greenhouse gases. Therefore, in terms of environmental impact, HSR is superior to automobiles and aircraft.

(1) Carbon monoxide (CO) emissions. Statistics show that the emission of carbon monoxide (CO) is 0.902 kg/person for the highway, 0.109 kg/person for the train, 635 kg/person for the aircraft and zero for the HSR as shown in Fig. 3.11.
    For HSR trains and cars, the amount of pollutants is shown in Table 3.8.
(2) Noise. Because HSR uses electric traction, there is no dust, soot and other exhaust gas pollution and the noise is lower than that of traditional rails (the

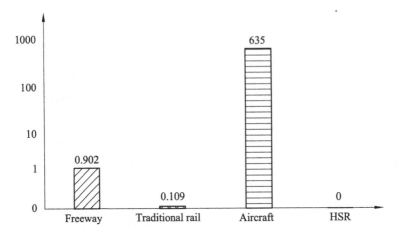

Fig. 3.11 CO emissions from different transportation (kg/person)

Table 3.8 Different pollutants of emissions from cars and HSR trains per person per kilometer	Pollutants	Pollution per person per kilometer/[g/(person/km)]	
		Cars	HSR trains
	CO	9.30	0.06
	NOx	1.70	0.43
	CH	1.10	0.03

noise generated by aviation per person per kilometer is 1, that of the bus is 0.2 and that of the HSR is 0.1) as shown in Fig. 3.12.

(3) Pollution control fees per person per kilometer. Statistics show that if the cost of pollution control per person per kilometer for HSR is 1, then the cost for the freeway is 3.76 and that for aviation is 5.21 as shown in Fig. 3.13.

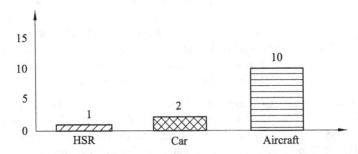

Fig. 3.12 Noise generated by different transportation per person per kilometer (Regard the noise generated by aviation per person per kilometer as 1)

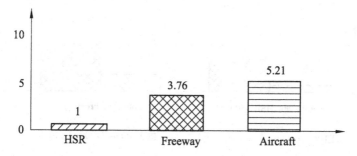

Fig. 3.13 Pollution control costs per person per kilometer for different transportation (Regard the cost of pollution control per person per kilometer for HSR as 1)

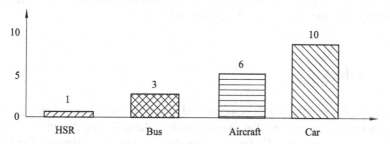

Fig. 3.14 Carbon emissions per person from different transportation (Regard carbon emissions of HSR as 1)

(4)   Carbon emissions per person. Statistics show that the carbon emissions per person of HSR are 1/10 of that of private cars, 1/3 of that of buses and 1/6 of that of airplanes as shown in Fig. 3.14.

3.7   Occupation of Land

Road transportation requires not only the construction of roads but also a large number of parking facilities occupying a large amount of land. At present, traffic congestion and parking difficulties in most cities are the drawbacks of road traffic while HSR does not have this problem. Generally, the 350 km/h roadbed base width of the two-lane HSR is 9.6–11 m, and the four-lane freeway pavement roadbed is 26 m wide. The double-lane rail land is 46,666.7 m^2/km and the four-lane freeway covers 70,000 m^2/km as shown in Fig. 3.15.

According to the comparative analysis in Fig. 3.15, the HSR covers only 2/3 of the area of the freeway but the passenger traffic per hour of the HSR is four times of that of the four-lane freeway. Compared with traditional rails, HSR is basically operated on the viaduct so many valuable land is saved. Compared with the aviation,

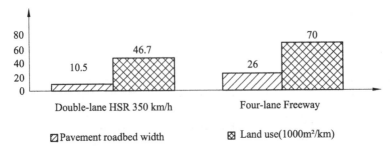

Fig. 3.15 Comparison of the land occupation of double-lane HSR and four-lane freeway

a large airport covers an area equivalent to a 1000 km double-lane HSR. In terms of land occupation, the advantages of HSR are obvious.

3.8 Comfort

The speed of HSR is lower than aircraft. However, in the short-distance journey (below 600 km), passengers can save much time because they do not need to go to the airport which is usually far away, nor do need check-in, baggage check and security check. What's more, the HSR has large space and passengers are more comfortable, as shown in Fig. 3.16.

The HSR train is very luxuriously arranged with complete working and living facilities, spacious and comfortable seats. It also has good running performance and very stable operation. Because of the shock absorption and sound insulation, the train is very quiet. Traveling on a HSR train is convenient and enjoyable. Statistics show that the sound of a mosquito is 40 dB, the noise in the cabin is about 81 dB when the

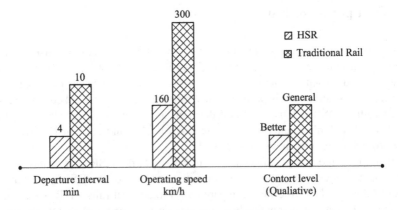

Fig. 3.16 Comparison of comfort between WHSR and traditional rail

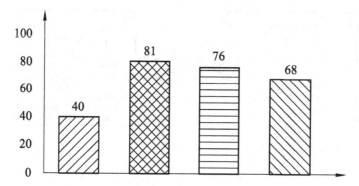

Fig. 3.17 Comparison of different vehicle noises

aircraft is flying, the noise of the car with a speed of 120 km/h is about 76 dB and the noise of the train of the Chinese HSR at the speed of 380 km/h is about 68 dB, as shown in Fig. 3.17.

3.9 Economic Benefits

HSR not only promotes the urbanization process of rural areas along the rail and creates new employment opportunities, but also restores investment quickly without causing fiscal and financial burdens. The construction cost of HSR is about 0.3–0.6 billion dollars per kilometer (China's HSR construction cost of 250 km/h is 0.8 billion Yuan/km, and 350 km/h of it is 130 million Yuan/km). But the industries driven by HSR radiation create 2.4 times value that of the construction cost.

Direct economic benefits. The released results show that when the HSR is extended forward, every 1 Yuan invested in line construction, vehicle manufacturing, power supply system, control system, station construction, etc. will stimulate the development of related industries and obtain a total practical value of 2.4 Yuan due to radiation reasons. For example, Japan's Tokaido Shinkansen had a total investment of 380 billion yen. Because the passengers has increased rapidly since it was put into operation and the transportation cost has been only one-fifth of that of the aircraft, the investment in the seventh year of official operation has been fully recovered. After 1985, the annual net profit reached 200 billion yen. Germany ICE HSR trains have an annual net profit of 1.07 billion marks. The French TGV annual net profit reaches 1.944 billion francs. China's HSR of Beijing—Shanghai line cost 208.84 billion Yuan and the total investment for a 10-year period was about 250 billion Yuan. Calculated according to 10 years of investment recovery, the annual dilute cost was 25 billion Yuan, and interest expenditure was about 29 billion Yuan. However, the cost was harvested in July 2015 and it began to earn profit. Along the HSR has become the most active and potential area of economic development. HSR will play an important role in supporting regional coordinated development, optimizing resource allocation

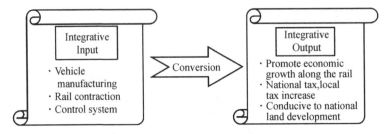

Fig. 3.18 Inputs and outputs of HSR

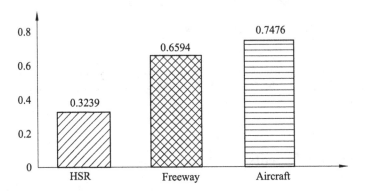

Fig. 3.19 Social costs of different vehicles

and industrial layout, building an efficient and comprehensive transportation system, reducing social logistics costs and promoting urban integration. Figure 3.18 shows the inputs and outputs of HSR.

Indirect economic benefits. In addition to direct economic benefits, HSR also has significant indirect social benefits. Statistics show that the social cost of the Beijing—Shanghai HSR is 0.3239 Yuan/(person·km), that of the freeway is 0.6594 Yuan/(person·km), and that of the aviation is 0.7476 Yuan/(person·km), with the ratio of 1:2.036:2.308. Under the condition of the same traffic volume, the social cost saved by building the Beijing—Shanghai HSR will reach 22.3 billion Yuan and the total operating amount in 6–7 years is equivalent to all construction investment. In addition, HSR can also drive economic growth along the lines and provide numerous employment opportunities. Figure 3.19 shows the social costs of different vehicles.

3.10   Social

HSR can not only reduce the distance between regions, but also promote the social development. On the one hand, the development of HSR can promote exchanges and cooperation between cities and promote the economic development of cities along the high-speed rail. On the other hand, the cities are formed by the HSR, which creates an economic circle and greatly promotes the development of society. HSR shortens the travel time, which is equivalent to narrowing the distance between regions, changing the concept of time and regional concepts. This can not only eliminate regional differences but also promote social equality, harmony and development. In particular, the space structure of the "four horizontal lines and four vertical lines" network of China's HSR is as shown in Fig. 3.20. The expansion of the HSR map and the increase of HSR lines have made the "high speed" "efficient" and "high quality" life a new normal.

The development of HSR has a major impact on national and regional development strategies. At present, countries all over the world have begun to build the transnational HSR in order to realize the rapid road passage between countries and regions as soon as possible, eliminate the influence of geographical location restrictions between countries and speed up economic and resource exchanges and cooperation between regions. From the overall national development perspective, the HSR has a far-reaching impact on the overall development of the country because it can guarantee national security at a strategic level. Considering the overall situation of global development, the development of HSR has a profound impact on the world political economy. HSR can promote world integration and realize "global village".

HSR development also promotes regional integration. This is because the good compatibility of the HSR is the basis for its international interconnection. Different types of HSR promote the integration of different regions. For example, WHSR promoted regional integration, MHSR promoted state integration, and SSR promoted global integration. Therefore, we must pay attention to the development of HSR.

3.11 Summary

Compared with other modes of transportation, HSR has advantages of big conveying capacity, fast speed, good safety, highly punctuality rate, low energy consumption, less impact on the environment, land conservation, comfort and convenience, considerable economic benefits and good social benefits. With its unique technical advantage to satisfy the demands for modern social and economic development, HSR has become an inevitable choice for the countries around the world. The development and operation of China's HSR shows that HSR has great development space and potential in China. China should make full use of its latecomer advantage to realize the leap-forward development of China's HSR.

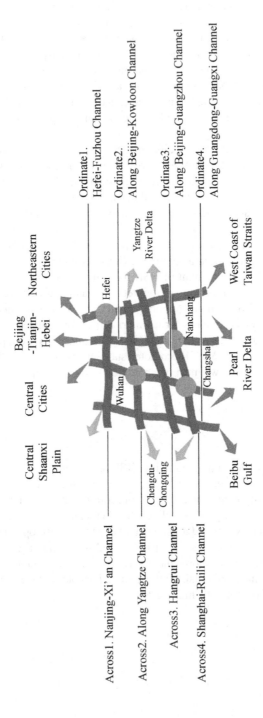

Fig. 3.20 HSR drives the economic circles

Chapter 4
Wheel High-Speed Rail (WHSR)

WHSR refers to the high-speed rail running on the track and operating at a speed of 200–400 km/h. The HSR, which people currently choose to travel, is also known as the conventional HSR. Currently, Japan, France, Germany and China are the four most mature countries in the development of WHSR technology. China is a country with the latest start of HSR, but the fastest development. By introducing HSR technologies from Japan, Germany and France, China has designed, through integration and innovation, eight types of high-speed rail trains, such as CRH1, CRH2, CRH3, CRH5, CRH380, CRH200, CRH300, CRH400, etc.. China has become the country with the longest mileage, the fastest operating speed of high-speed rail and the largest network of lines in the world, as shown in the Fig. 4.1.

HSR is the innovation and integration of many advanced science and technology. It is also the embodiment of the level of science and technology in various countries. At present, the main WHSR technologies in the world include TGV in France, ICE in Germany, Shinkansen in Japan and CRH in China. Train technology for WHSR in various countries is as shown in Table 4.1.

European countries mainly use the French TVG technology. Among Asian countries and regions, South Korea adopts French technology. Taiwan (a province of China) HSR introduces Japan's Shinkansen train technology, but the power protection system adopts European standard; The CRH express train in mainland China is an innovation based on the integration of the train technology of ICE, TVG of France and Shinkansen of Japan. The comparison of technical standards of WHSR trains in Japan, France and Germany is as shown in Table 4.2.

At present, French TVG technology is used in seven countries all over the world. It is the most widely used HSR technology in the world. Figure 4.2 shows the technical relationship of WHSR in the world.

(1) Japan's WHSR. Japan built the first HSR in the world, so it also pioneered the concept of high-speed ground transportation. The first Tokyo—Osaka Shinkansen line was opened in 1964, operating at a speed of 210 km/h, and the rail is later extended to 1079 km. The improvement and extension of the Shinkansen line have been ongoing, and the operating speed has increased to

© Southwest Jiaotong University Press 2023
Q. Hu and S. Qu, *A Brief History of High-Speed Rail*,
https://doi.org/10.1007/978-981-19-3635-7_4

Fig. 4.1 China's WHSR (CRH)

Table 4.1 Technical list of WHSR in various countries of the world

Countries and regions		Technology	Introduction of foreign talents
China	Mainland	CRH	Japan, Germany, France
	Taiwan	Introduction of Japanese Shinkansen system	Japan
Russia		Study on the Velaro RUS EVS to Russian wide rail on the basis of ICE3	Germany
Korea		KTX, Adopt type TGV-A train from France	French
Japan		Shinkansen	Japan
French		TGV	French
England		Adopt TGV Technology from French, The train is called Eurostar	French
Germany		ICE	Germany
Spain		Adopt TGV in the early stage, and adopt TALGO (Spain) and ICE (Germany) later	French, Germany
Italy		ETR train, Pendolino	Italy

240 km/h and 270 km/h with the improvement of the vehicle structure, and eventually reached 300 km/h. Lines with speeds above 350 km/h are also being designed. These lines are highly reliable, comfortable and safe and there has been no passenger death in decades of operation, as shown in Fig. 4.3.

(2) French WHSR. France opened its first HSR line from Paris to Lyon in 1981 as shown in Fig. 4.4. The line attracted many passengers from the start, and it ran at a speed of 270 km/h. The Atlantic line from Paris to southeastern France, the line from Paris to northern Lille, and the Channel Tunnel were later built.

Table 4.2 Comparison of technical standards of WHSR trains in Japan, France and Germany

Index	Japan			France		Germany
	Dokaido Shinkansen	Sanyo Shinkansen	Joetsu Shinkansen	Southeast line	Atlantic line	ICE
Line length/m	515	554	766	410	284	426
Construction period/year	1959–1964	1967–1975	1971–1991	1976–1983	1985–1990	1976–1991
Design maximum speed/(km/h)	210	260	260	360	350	300
Maximum operating speed/(km/h)	270	270–300	275	270	300	280
Minimum radius of curvature/m	2500	4000	4000	4000	6250	4670
Minimum longitudinal curvature/%	10,000	15,000	15,000	25,000	25,000	22,000
Steepest slope/%	20	15	15	35	25	12.5
Rail center distance/m	4.2	4.3	4.3	4.2	4.2	4.7
Width over sides of car body/m	3.4	3.4	3.4	2.9	2.9	3.1
Construction roadbed surface width/m	10.7	11.6	11.4	13.6	13.6	13.7
Cross section area of double track tunnel/m^2	64	64	64	NA	71	82

The line from Lyon along the Mediterranean to Marseille opened in 2001 with a maximum operating speed of more than 330 km/h, as shown in Fig. 4.4.

(3)   Germanic WHSR. Germany opened its first HSR in 1991 as shown in Fig. 4.5. The line from Hanover to Fort Ulrich has a maximum speed of 250 km/h. Germany is a pioneer in speed-up of existing lines, many of which have been raised to 200 km/h with less investment than new lines. Several new lines have been built or opened in Germany including Mannheim Stuttgart, Frankfurt Cologne, Berlin-Hanover, and Berlin-Hamburg. Figure 4.5 shows the Germanic WHSR.

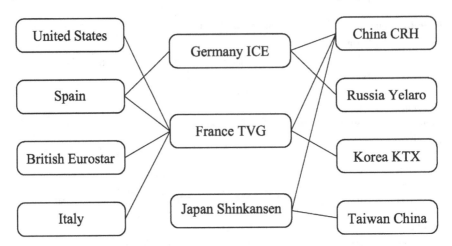

Fig. 4.2 Technical relationship of WHSR in the world

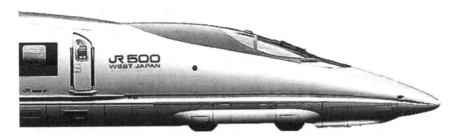

Fig. 4.3 Japan's WHSR

Fig. 4.4 French WHSR

(4)   China's WHSR. China opened its first HSR in 2008, an inter-city train from Beijing to Tianjin. Later, the Wuhan—Guangzhou section of the Beijing—Guangzhou HSR was officially launched on December 26, 2009, with a maximum operating speed of 350 km/h, and the travel time from Wuhan to Guangzhou was shortened from about 11 h to about 3 h. The travel time from Wuhan to Changsha is only 1 h, and the travel time from Changsha

Fig. 4.5 Germanic WHSR

Fig. 4.6 China's WHSR trains

to Guangzhou is only 2 h. Wuhan—Guangzhou HSR has become the fastest and densest HSR around the world. At present, on the basis of perfecting the HSR network of "four-vertical lines and four-horizontal lines", China has also planned the HSR network of "eight vertical lines and the eight horizontal lines". Figure 4.6 shows China's WHSR trains.

4.1 Basic Characteristics of WHSR

The highest operating speed is the most important index of HSR. At present, the highest operating speed of WHSR is 250–350 km/h. France has recently set a record of running 1000 km at an average speed of 317 km/h, while China has set a record of running more than 10,000 km at an average speed of 350 km/h.

4.1.1 Particularity of WHSR

With the improvement of train speed, the construction standard of WHSR has been raised. Especially, the HSR train will have very high request to the cross-section of the tunnel when it passes through the tunnel. Therefore, WHSR has its particularity including: space problem, curve problem and tunnel problem. Figure 4.7 shows the structure of WHSR.

(1)   The space problems of WHSR. When a HSR train runs along the ground at high speed, it will drive the air around the train and form a specific train wind. When the two trains of adjacent lines meet each other at the same high speed, the shock wave of air pressure is easy to shatter the window glass, which makes the passenger's ear uncomfortable, and even affects the smooth running of the train. Therefore, the track of WHSR requires a wide range of driving space, as

Fig. 4.7 Structure of WHSR

Fig. 4.8 Driving space of HSR

shown in Fig. 4.8. It can be solved by increasing the distance between the two lines and widening the safe retreat distance of passengers on the platform.

(2) Curve problems of WHSR. WHSR has put forward higher technical requirements for the route curve, as shown in Fig. 4.9. The high smoothness of the HSR track requires that the spatial route curve be as gentle as possible, that is, the variation of the horizontal and vertical section of the track is as smooth as possible. At the same time, the centrifugal acceleration generated by the train

Fig. 4.9 Curve design of HSR

Fig. 4.10 Passing through tunnel of HSR

running on the curve is proportional to the square of the train speed, which directly affects the comfort, stability and safety of the train. Therefore, the higher the driving speed is, the bigger the increase of the radius of plane curve and elevation curve is.

(3) Tunnel problems of WHSR. The tunnel of WHSR is larger than that of traditional rail, the force is more complicated, and the speed of HSR train is higher than that of ordinary train, and the tunnel maintenance has certain time limit. The requirements of safety line, durability and waterproof performance of tunnel lining are improved, as shown in Fig. 4.10. In addition, the strength of the bottom of the tunnel is higher than that of the traditional rail, and the cross-section span of the tunnel is larger, so the requirements of the thickness of the bottom slab and the strength of concrete are also high.

4.1.2 The Difference Between WHSR and Traditional Rail

The running speed of WHSR is faster than that of traditional rail. Therefore, in order to meet the requirement of higher speed, the technical performance and track requirements of WHSR are different from those of traditional rail. Figure 4.11 shows the WHSR and traditional rail.

(1) WHSR is running smoothly to ensure the safety and comfort of the train. The WHSR adopts seamless rail, and the WHSR with a speed of more than 300 km/h adopts ballast less track, which is an integral track bed without stone to ensure smooth ride (Fig. 4.12).

(2) WHSR rail has few bends and a large radius. At present, WHSR basically goes straight on the elevated frame (Fig. 4.13).

(3) Elevated bridges and tunnels are widely used in WHSR to ensure riding comfort and shorten distance.

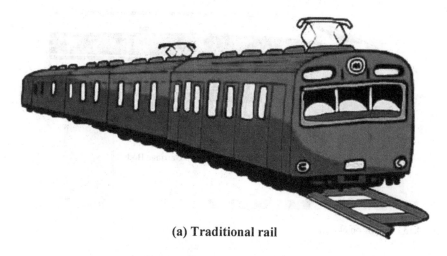

(a) Traditional rail

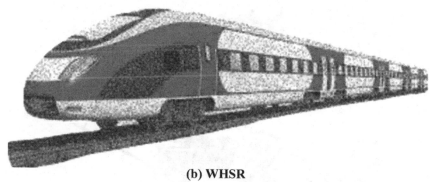

(b) WHSR

Fig. 4.11 WHSR and traditional rail

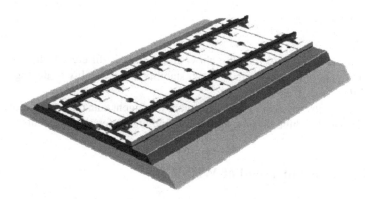

Fig. 4.12 Ballastless track

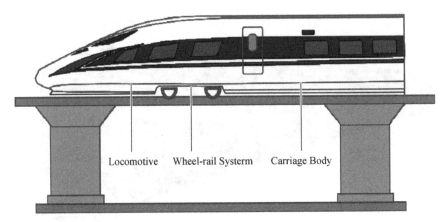

Locomotive   Wheel-rail Systerm   Carriage Body

Fig. 4.13 WHSR on elevated

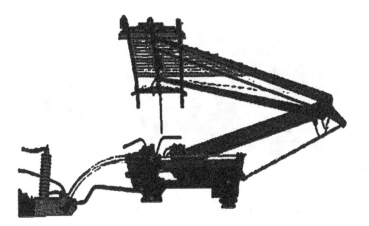

Fig. 4.14 WHSR pantograph

(4) Suspension mode of the wire above the WHSR is different from that of the traditional rail to ensure the contact stability and durability of the high-speed EMU (Fig. 4.14).

(5) Signal control system of WHSR is more intelligent than that of traditional rail, because of its high density, high speed and high safety requirements.

4.2 Development Trend of WHSR

In view of the considerable economic benefits and immeasurable political influence of the HSR, many countries in the world have invested in the construction process of HSR. Due to the maturity of WHSR technology, many countries and regions in

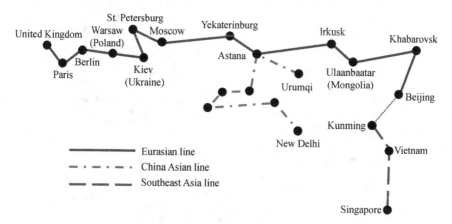

Fig. 4.15 Sketch map of the forward planning of China's WHSR

the world have opened WHSR and many countries and regions are building and planning WHSR. Figure 4.15 shows the Sketch map of the forward planning of China's WHSR.

4.2.1 Development Trend of WHSR in Europe

The Council of Europe was convened in Germany in 1994, decided to implement the resolution on the construction and expansion of the Trans-European transport network. In 1998, the International Railway Union began to organize and study further plans for the European highway network, and called for the formation of European HSR network by 2020, as shown in Fig. 4.16.

The 10 WHSR lines in Europe are:

① Helsinki—Tallinn—Riga—Vilnius—Warsaw (Gudansk);
② Berlin—Poznan—Warsaw—Minsk—Moscow—Lower Novgorod (eastward to connect Siberian Rail);
③ Berlin (Dresden)—Katowi—Kiev;
④ Nuremberg   (Dresden)—Bragg—Bratislava—Budapest—Bucharest—Constantine (Istanbul);
⑤ Venice—Cobo—Riquega—Budapest—Uz Grode—Volvot;
⑥ Gudansk Poznan Catoves—Vienna;
⑦ Vienna—Budapest—Belgrade—Bucharest—Odessa;
⑧ Tirana—Skopuki—Sophia—Levarna;
⑨ Helsinki—St.   Petersburg—Minsk—Kaliningrad—Kiev—Alexandria—Odessa;
⑩ Salzburg—Zagreb—Belgrade—Sofia—Athens.

Fig. 4.16 European HSR network topological graph

Development Trend of WHSR in Germany

German HSR technology developed earlier than other countries. In 1988, their electric traction train's test speed exceeded 400 km/h, reaching 406.9 km/h. However,
the construction of practical HSR in Germany did not begin until the beginning
of the 1990s, due to the dense cities in Germany where passenger traffic was most
concentrated, and the HSR was developed and perfected. The construction of another
HSR obviously could not serve the purpose of attracting passenger. The ICE intercity high-speed train ran on the HSR line with the speed of 250 km/h. In 1993,
the ICE HSR train entered Berlin, and since then, the Germany capital has been
included in the ICE rapid transit system. In addition, ICE also crossed the border
between Germany and Switzerland achieving the direct international transportation
from Zurich to Frankfurt.

(1) Operation state of HSR. At present, Germany has two HSR lines. The one
      is Mannheim—Stuttgart line completed in 1991, and the other is Hanover—
      Wirzburg line opened in 1992.

①   Line characteristics. The intercity express lines that Germany is operating are Hannover Fulga Wirzburg, Manheim Stuttgart, Hanover Berlin, Cologne Frankfurt, Cologne Dylan, Rastatnam Orfenberg, Hamburg Berlin, Nuremberg—Ingolstadt, Munich—Augsburg, etc., total mileage over 2000 km, as shown in Table 4.3.

     As far as the distribution of HSR lines between cities in Germany, German HSR lines have not yet formed a network. HSR lines are spread into three parts in space, connecting the central Germany, the eastern and western Germany and the interior of the southeast of Germany, respectively. All parts of the lines are still connected by conventional trains with speeds below 200 km/h, which is shown in Fig. 4.17.

②   Technical characteristics. There was only one serious train derailment incident since Germanic fast intercity train operated in 1991, which caused 101 deaths and 88 serious injuries. The survey results indicated that the train derailment happened due to defective material used in the wheel. This accident affected the development plan of HSR all over the world. Germany slowed down the construction plan for fast intercity train and focused on train's safety performance. Germany also adopted the technological path of independent development in the HSR field. In 1985, Germany exploited their first ICE-V testing train which can run 409.6 km/h. At present, the ICE-5 is adopted magnetic suspension technique, and the stability and safety performance have all been improved. In addition, the interior layout design is more user-friendly (Table 4.4).

(2)   The development trend of high-speed rail. According to the HSR project in Germany, the four intercity Express lines of Frankfurt—Mannheim, Stuttgart—Ulm, Hamburg—Hanover and Seelze—Minden are under planning and construction. Among them, the line of Frankfurt-Mannheim will connect the completely eastern region to western region of Germany. Besides that, Germany realized the intercity express transportation networks with surrounding countries, and they become the kernel trunk line to connect Europe. Now, Germany used the lines of Kolner—Düren, Offenburg—Basel, Munich—Augsburg and Berlin—Poznan, etc. for connecting Belgium, France, Switzerland, Austria and Poland respectively and realizing networked of HSR in Europe.

Development Trend of WHSR in France

France is one of the earliest countries to develop HSR in Europe. French TGV HSR technology has been the most widely used HSR technology in the world. At the same time, the TGV has the fastest average speed of wheel express train over the world. The initial success of TGV promoted the expansion of rail networks, and many new rail lines has been built in South, West and Northeast of France. The TGV connects Switzerland through France rail networks, connects Belgium, Germany and Netherlands trough the Northwest railway networks of France, and connects Great

Table 4.3 Comprehensive present situation of HSR lines in Germany

Status	Line name	Origin station	Terminal station	Mileage/km	Operating speed/(km/h)	Operation year
Operating	Hanover—Wirzburg	Hanover	Wirzburg	327	250	1991
	Mannheim—Stuttgart	Mannheim	Stuttgart	99	250	1991
	Hanover—Berlin	Hanover	Berlin	148	250	1998
	Kolner—Frankfurt	Kolner	Frankfurt	219	300	2002
	Kolner—Düren	Kolner	Düren	39	250	2003
	South Raststätte—Offenburg	South Raststätte	Offenburg	44	250	2004
	Hamburg—Berlin	Hamburg	Berlin	286	230	2004
	Nuremberg—Ingolstadt	Nuremberg	Ingolstadt	89	300	2006
	Munich—Augsburg	Munich	Augsburg	43	230	2010
	Grobers—Erfurt	Grobers	Erfurt	123	250	2015
	Nuremberg—Erfurt	Nuremberg	Erfurt	190	250	2016
	Offenburg—Basel	Offenburg	Basel	121	250	2015
Planning	Frankfurt—Mannheim	Frankfurt	Mannheim	85	300	2017
	Stuttgart—Ulm	Stuttgart	Ulm	84.8	250	2019
	Hamburg—Hanover	Hamburg	Hanover	114	300	Pending
	Seelze—Minden	Seelze	Minden	71	230	Pending

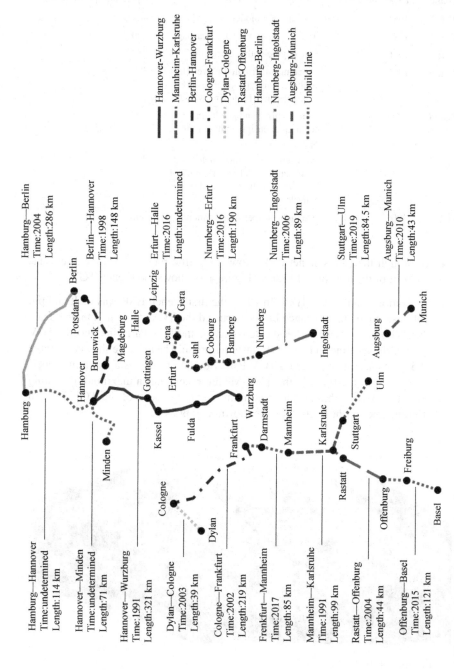

Fig. 4.17 Network of HSR operation lines in Germany

Table 4.4 List of the development of Germanic train technology

Name	Experiment time	Technical characteristics	Operating speed/(km/h)
ICE-V	1985	Testing train model	409.6
ICE-1	1991	Two engines with 10–12 carriages	280
ICE-2	1996	One engine with 7 carriages	280
ICE-3	1997	It belongs distributed-power electric multiple unit type, totally has 8 carriages and, 16 driving wheels	300
ICE-T	1998	Electric drive and pendulum technique, design without running speed	200
ICE-4	2004	Non-separated train, it belongs the distributed-power electric multiple unit type, totally has 120 driving wheels	300
ICE-5	2005	Magnetic suspension technique	400

Britain through the Eurostar network. Now, the traffic range of TGV express has covered most of the territory of France. Figure 4.18 shows the French TGV.

(1) Operation condition of HSR. In 1981, the first French HSR was built and put into use. Since then, France has constructed the TGV Atlantic line, the TGV North line, the TGV Southeast extension line, the TGV Paris link line, the TGV Mediterranean line and the TGV Eastern Europe. The lines are showed in Table 4.5. The earliest constructed TGV Paris Southeast line has a maximum operating speed of 270 km/h, the TGV Mediterranean has a maximum operating speed of 300 km/h, and the maximum operating speed of the TGV Eastern Europe line reaches 320 km/h, as shown in Table 4.5.

Fig. 4.18 French TGV

Table 4.5 Operating lines in France

Line name	Distance/km	Maximum speed/(km/h)	Operating time
Southeast line	409	270	1981
Atlantic line	282	300	1989
North line	333	300	1993
Southeast extension line	148	300	1992
Paris eastern link line	128	300	1994
Mediterranean line	251	300	2001
East line (first phase)	300	320	2007
Rhine-Rhone line (East segment)	140	320	2011
Perpignan-Figueres line	52.7	320	2009
Bretagne-Loire region line	180	320	2012
South Europe-Atlantic line	120	320	2013
South Europe-South Atlantic line	120	320	2013
Poitiers-Limoges line	115	320	2015
East line (second phase)	106	320	2015
Bordeaux-Toulouse line	230	320	2016
South Europe-North Atlantic line	180	320	2016

(2)　Development trend of HSR. The network of French HSR now has formed the major structure of the HSR network, which was centered with Paris and radiated in all directions to the southeast and northwest. There are outer ring lines in the Paris and Lyon regions, and the trains of North line can be routed through the Paris outer ring road to the Southeast line, which are shown in Fig. 4.19.

With the rapid development of HSR over the world, France has its own new rail planning. It includes 8 lines of the Rhine—Rhone line and the Lyons—Turin line, etc. which are showed in Table 4.6. It is estimated that by 2020, France would have built a high-speed line with a total mileage of more than 700 km, as shown in Table 4.6.

Development Trend of WHSR in Italy

Italy built the first HSR line from Rome to Florence in 1992. After two years of research, it officially began to construct the HSR network in 1994. In 1998, Italy upgraded the section of 180 km rail of Milan—Bologna line, and then the train speed was increased to 300 km/h. After 2000, the HSR lines such as Turin—Bologna, Milan—Venice and Milan-Genoa were opened continuously.

(1)　Operation condition of HSR. Italy is one of the first countries in Europe to build HSR line. The HSR line of Rome—Florence with the distance of 254 km was

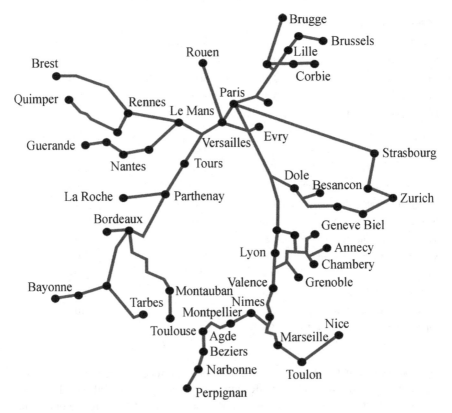

Fig. 4.19 French HSR network

Table 4.6 Planned HSR in France

Line name	Distance/km	Operating time
Rhine—Rhone line	185	2022
Lyons—Turin new link line	150	2020
Provence—Alps—Côte d'Azur area line	200	2020
Montpelier—Perpignan new line	150	2022
The Southern link line of Paris	40	2020
Bordeaux—Spain line	230	2020
Picardie region line	250	2020
Paris—Lyons second line	430	2025

built and put into operation in 1992. In the mid-1990s, Italy further planned to build the "T" shaped high-speed transport network, which crossed the East and West and threaded the North and the South. However, affected by political and economic factors, the plan was delayed. In 2005–2006, Italian government adjusted its national transportation policy and decided to invest in accelerating the construction of the North–South HSR transportation corridor, and built a new high-speed line that connecting the major cities in the North and the South and connected with the completed HSR Direttissima in sections.

With the opening of the new HSR line in Italy, the main cities along the line, such as Bologna, Florence and Rome, have been linked together, forming a HSR network with a total length of more than 10,000 km, as shown in Fig. 4.20. The Italian HSR program is part of the EU HSR project and will be directly operational with HSR trains in neighboring countries such as France, Switzerland and Slovenia in the future.

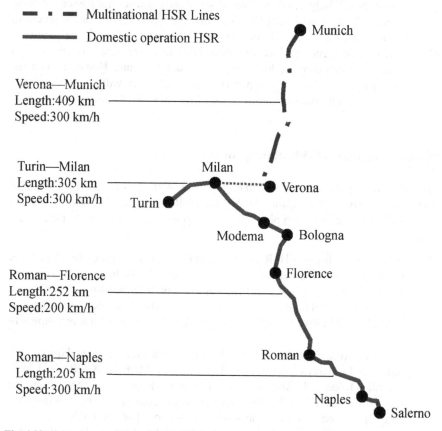

Fig. 4.20 Operation condition of Italian HSR

Table 4.7 Operation lines of Italian HSR

Origin station	Terminal station	Mileage/km	Operating speed/(km/h)	Operating time
Rome	Florence	252	200	1992
Rome	Naples	205	300	2005
Verona	Munich	409	300	2007
Turin	Salerno	918	300	2009
Turin	Milan	305	300	2009

Italian HSR has been in operation for more than 20 years, and there have been security incidents during this period. On April 26, 2012, two Italian HSR trains "Frecciarossa" crashed accidentally when they came into Rome Termini Train Station, this accident made 6 people light injured. On September 24, there happened an accident of a HSR train collided with a truck in suburb of Chestell Nino of South Italy, the train diver died instantly and several passengers were injured. The two accidents have sounded the alarm for the development of Italian HSR, but in general, the operation of Italian HSR is relatively safe.

(2) Development trend of HSR. European HSR lines are generally centered on France and Germany, followed by Spain and Belgium. However, with the completion of Italian "T"-shaped HSR network, Italy will also become one of the most important countries in the HSR network of Europe, as shown in Table 4.7.

Development Trend of WHSR in Spain

Spanish HSR (Spanish: Alta Velocidad Española, AVE) has a maximum speed of 300 km/h. In April 1992, Spain operated the HSR line from Madrid to Seville on the eve of the Barcelona Olympics and caught up with the development pace of the world.

(1) Operation condition of HSR. After its first HSR line was successfully operated, Spain continued to accelerate the development of HSR trains and formulated new road network plans. After the construction and reconstruction, Spanish HSR has formed a modern freeway network and become one of the countries with advanced railways. Spain's existing HSR operation lines are shown in Table 4.8.

The existing railway network in Spain is a broad rail line, and 3 quasi track high speed railways have been built as follows: Madrid—Seville (471 km), Madrid—Toledo (74 km), Madrid—Lleida (481 km). The above lines are equipped with ETCS/ERTMS-1 train operation control system, running AVE series HSR trains, with the maximum running speed of 350 km/h.

Spanish HSR train (AVE) belongs to Spanish national rails. At that time, Spain was preparing for constructing a new rail to connect Madrid and Andalusia. At the same time, Spain was also considering whether to build

Table 4.8 Spanish existing HSR operation lines

Number	Origin station	Terminal station	Speed/(km/h)	Distance/km
1.	Madrid	Seville	250	471
2.	Madrid	Barcelona	280	630
3.	Madrid	Toledo	250	74
4.	Madrid	Valencia	250	438
5.	Barcelona	Valencia	220	389
6.	Valencia	Alicante	280	160
7.	Madrid	Tarragona	250	550

it into HSR, in order to connect to the big cities of Barcelona and Valencia and in the future to promote their sustainable development. At 8:42 pm, on July 24, 2013, the Spanish HSR train numbered Alvia 151 derailed on the way from Madrid to the Northern city of Fello, causing 77 deaths and 131 injuries. This is one of the biggest accidents in the history of Spanish rails, but Spain still has some lessons to learn in terms of train safety and technology.

(2)   Development trend of HSR. Currently the HSR lines being built and planned by Spain are: Madrid—Barcelona—Southwestern of France, Zaragoza—Bilbao, Logroñok—Southwestern of France, Madrid—Lisbon (the capital of Portugal). The old lines which will be transformed include the line of Madrid—Valencia, Madrid—Leon, Valladolid—Logroño, Seville—Verfa and Seville—Cadiz, etc. The main HSR trains in Spain are: Alaris, Alvia, Anant and Eurome.

By 2020, Spain will build a HSR line of 10,000 km. By then, 90% of Spanish nationals will encounter a HSR station every less than 50 km. Spain government will focus on building HSR lines to the North and the Mediterranean coast on the West coast of the Atlantic, and connecting the main populated areas to form a radial HSR network, as shown in Fig. 4.21.

Development Trend of WHSR in UK

The UK is the earliest country in the world to build the rail. It developed the first steam-type rail train as early as 1804. The British Railway has developed from a small and decentralized company into the four major rail giants which now are controlled by the British Rail Transport Department. It achieved the profitability of the railway through internal combustion and electrification. By the beginning of the twenty-first century, the total length of British railways has reached 1.66 million km.

(1)   Operation condition of HSR. The development of HSR in UK is relatively slow, and so far only one HSR line has been opened. High-speed Rail Line 1 (HS1) is the London-Paris line, the route passes the English Channel and connects France and the United Kingdom. At present, the British High-speed Rail Line 2 (HS2) is under construction, see Fig. 4.22. The construction of the

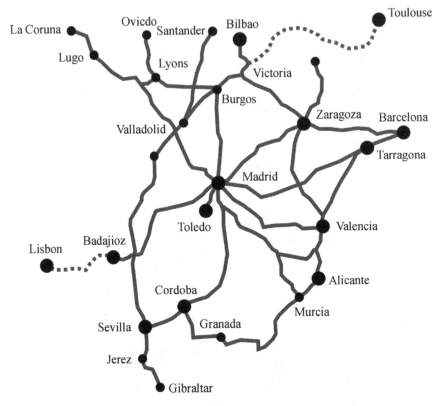

Fig. 4.21 Lines of Spanish HSR

HS1 took a full 16 years, whose total mileage was 109 km. Moreover, in 2003, at the beginning of the opening of the HSR, it only maintained a high-speed of 300 km/h in the French section, but when entered the British section, still maintained the maximum speed limit of 160 km/h for the traditional rails.

(2)   Development trend of HSR. It was not until 2007 that UK sped up the HS1 route and really realized the full speed of Paris to London. British HS2 runs through the North and South of UK, with a total length of 525 km. It travels from London in the South to Birmingham in the middle of England, where it splits into the Northern cities of Leeds and Manchester. After the opening of HS2, the time from London to Manchester is nearly half shorter. The HS2 HSR project was carried out in two phases. The first phase of the project, from London to Birmingham, was 225 km in length. It was launched in 2012, started in 2017 and expected to be completed in 2026. The second phase of the project, from Birmingham to the northern cities of Leeds and Manchester, was 300 km length. The construction plan was announced in 2013 and was expected to start in 2026 and be completed in 2033. In other words, the construction of HS2's entire line will take 20 years from start-up to completion. The first phase of

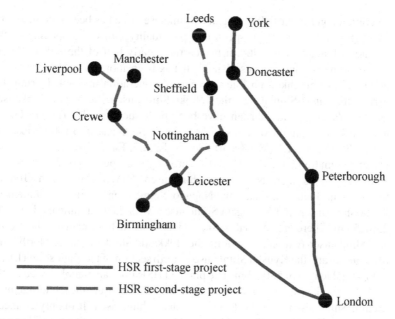

Fig. 4.22 British HSR

the HS2 project will reduce the current time from London to Birmingham by half an hour.

4.2.2 Development Trend of WHSR in Asia

Asia is the largest continent in the world regardless of population or land area. However, not many countries in Asia have the WHSR, mainly are China, Japan, South Korea and Turkey.

Development Trend of WHSR in Japan

Japan's transportation is developed, whether the perfection of water transport, aviation or road and rail systems are in the leading position in the world. Especially the Shinkansen express train technology, which was first developed in Japan after the 1960s, has become the symbol of the world's advanced transportation. Japanese rail transportation network consists of national rails across the country, private rails between large and medium-sized cities, and subways in the city. So far, Japan's rail operating lines have exceeded 22,000 km, of which the Shinkansen operating length has reached 2501.5 km, and the total number of passengers of Japan's eight Shinkansen lines in operation has reached 89 billion.

(1)   Operation condition of HSR. Japan's Shinkansen line has been recognized by passengers for its comfort, speed and punctuality. Since the opening of the Tokaido Shinkansen in 1964, it has carried nearly half of the rail passenger flow volume on the Japanese island. It has been more than 50 years since the Tokaido Shinghkansen line was put into operation in 1964. During this period, the Sanyo Shinkansen, the Tohoku Shinkansen, the Joetsu Shinkansen and the Yamagata Shinkansen have been built successively. The HSR train has developed from the 0-series of the Tokaido Shinkansen to 100 Series, 200 Series, 300 Series, E1 (MAX), 400 Series, E2, E3, E4, and E5 series, etc.

(2)   Development trend of HSR. The HSR currently was operated by Japan include eight HSR lines such as Tokaido Shinkansen, Sanyo Shinkansen, Tohoku Shinkansen, Joetsu Shinkansen, Nagano Shinkansen, Kyushu Shinkansen, Akita Shinkansen and Yamagata Shinkansen, with the total mileage length of 2501.5 km. There are four HSR lines under construction, namely, the Northeast Shinkansen (extended section), the Hokkaido Shinkansen, the North Land Shinkansen and the Kyushu Shinkansen (north section of the Kagoshima Line), with a total length of 580.5 km. The planned HSR lines are mainly based on the Shikoku Shinkansen, the Shikoku intersection Shinkansen, the Chūgoku intersection Shinkansen, and the Kyushu Transit Shinkansen. It mainly connects with the HSR of China, North Korea and South Korea, forming a rapid ground passenger transport corridor as shown in Table 4.9. The four rails currently under construction in Japan will be put into use successively. Figure 4.23 shows the overview of Japan's HSR operating lines.

From the technical level, HSR can be divided into HSR ticket inspection technology, HSR train design technology, and HSR track protection technology. Japan's safety research in these three areas has been at leading level in the world. Japan's self-developed EMU from the 0 series to E5 series have greater advantages in terms of stability and speed than other countries. There are 13 types of the HSR trains currently operating in Japan like Nozomi number, Hikari number, and Kodama, Kotama, Hayate, Yamagam, Nasuye, Toki/Tanigawa, Asama, Tsubame, Komachi, and Tsubasa, they are responsible for passenger flow transport task of different routes.

Development Trend of WHSR in Korea

Since its privatization on January 1, 2005, the Korean rail has been operated by the Korea National Railways Corporation, which is headquartered in Seoul. There are 15 lines across the country with a total operating mileage of more than 3472 km, of which the length of HSR line is 420 km.

(1)   Operation condition of HSR. On April 1, 2004, the first "Gyeongbu" HSR line from Seoul to Busan (Daegu) was completed and opened to traffic. The maximum speed of the train was 300 km/h. In December of the same year, one group of Korean HSR-350X trains was tested. In the test, the speed was 352.4 km/h. At present, the Korea High-speed Rail (KTX) is operated by the

Table 4.9 Comprehensive present situation of Japan's HSR lines

Condition	Line name	Origin station	Terminal station	Mileage/(km)	Operating speed/(km/h)
Operating	Tokaido Shinkansen	Tokyo Station	Shin—Osaka	515.4	300
	Sanyo Shinkansen	Shin—Osaka Station	Hakata Station	553.7	300
	Tohoku Shinkansen	Tokyo Station	Hachinohe Station	631.9	275
	Joetsu Shinkansen	Miya Station	Niigata Station	269.5	240
	Nagano Shinkansen	Takasaki Station	Nagano Station	117.4	275
	Kyushu Shinkansen	Shin—Yatsushiro Station	Kagoshimachuo Station	137.6	260
	Akita Shinkansen	Morioka Station	Akita Station	127.3	130
	Yamagata Shinkansen	Fukushima Station	Shinjō Station	148.6	130
Constructing	Tohoku extension	Hachinohe Station	Shin—Aomori Station	81.2	About 300
	Hokkaido Shinkansen	Shin—Aomori Station	Hokkaido Station	148.8	About 300
	Hokuriku Shinkansen	Nagano Station	Kanazawa Station	220.6	About 300
	Kyushu Shinkansen	Hakata Station	Shin—Yatsushiro Station	129.9	About 300
Planning	Kyushu extension	Shin-Tosu Station	Nagasaki Station	129.9	About 300
	Hokuriku extension	Kanazawa Station	Shin—Osaka Station	254	About 300
	Hokkaido extension	Shin—Hakodate Station	Sapporo Station	211.5	About 300
	Shikoku Shinkansen	Shin—Osaka Station	Matsuyama	480	About 300
	Shikoku intersection Shinkansen	Okayama Station	Kochishi	About 160	About 300
Planning	Chūgoku intersection Shinkansen	Okayama Station	Matsue	About 180	About 300
	Kyushu intersection Shinkansen	Oita	Kumamoto	About 118	About 300

(continued)

Table 4.9 (continued)

Condition	Line name	Origin station	Terminal station	Mileage/(km)	Operating speed/(km/h)
	Sanin Shinkansen	Shin—Osaka Station	Shimonoseki	About 505	About 300
	Uetsu Shinkansen	Toyama	Shin—Aomori Station	560	About 300
	Ohu Shinkansen	Yamagata	Akitashi	270	About 300
	Hokuriku Chukyo Shinkansen	Tsuruga	Nagoya	50	About 300
	Centre Shinkansen	Tokyo Prefecture	Osaka	About 490	About 300

Fig. 4.23 Overview of Japan's HSR operating lines

Fig. 4.24 Korean WHSR train

Korean Railway (Korail) with a total mileage of 420 km. The vehicles are equipped with French TGV technology. The maximum speed can reach more than 300 km/h. Figure 4.24 shows the Korean WHSR train.

(2) Development trend of HSR. South Korea's HSR from Seoul to Busan has been completed. By using this HSR line, as well as existing rail lines, South Korea's HSR runs two lines: Gyeongbu Line (Seoul—Busan): (Haengsin—) Seoul—Yongsan—Gwangmyeong—CheonanAsan—Daejeon—DongDaegu—Milyang—Gupo—Busan; HonamLine (Seoul—Mokpo): (Haengsin) Seoul—Yongsan—Gwangmyeong—CheonanAsan—Seodaejeon—Nonsan—Nonsan—Iksan—Gimje—Jeongeup—Jangseong—Kwangju/Naju·Mokpo. In addition, Korea is currently planning one HSR connecting major cities to form a "three vertical and two horizontal" structure in space as shown in Table 4.10.

South Korea's plan of HSR routes are divided into fast-speed and high-speed according to line conditions. Here "fast-speed" refers to a line with a train operating speed of 230 km/h, and "high-speed" refers to the line whose train operating speed is set to 250 km/h. Currently, Korea is planning the lines of Wonju—Jangtu, Wonju—Jecheon, Hattan—Yeongcheon, Dongtan—Hongseong as high-speed lines, others are fast-speed lines, as shown in Fig. 4.25.

Development Trend of WHSR in China

On August 1, 2008, the Beijing—Tianjin intercity rail with a speed of 350 km/h was put into operation, which was a major milestone in the history of China's railway development. From December 2009 to October 2010, Wuhan—Guangzhou, Zhengzhou—Xi'an, Shanghai—Nanjing and Shanghai—Hangzhou high-speed rail were opened and operated successively; In particular, in July 2011, the Beijing—Shanghai high-speed rail was opened and put into operation, and the CRH380 series

Table 4.10 Route planning of Korean HSR

Route planning	Line name	Via Major cities	Distance/km
Three vertical	Sokcho—New Gyeongju	Sokcho—Kangnung—Sandou—Pohang—Singyeongju	About 280
	Seoul—Busan	Seoul—Wonju—Jecheon—Hattan—Yeongcheon—Dongdaegu—Samnangjin—Busan	420
	Hongseong—Yeosu	Hongseong—SinChanghang—Iksan—Jeonju—Suncheon—Yeosu	About 271
Two horizontal	Incheon Airport—Chuncheon	Incheon Airport—Seoul—Chuncheon	About 102
	Mokpo—Busan	Mokpo—Kwangju—Suncheon—Jinju—Masan—Busan	About 247

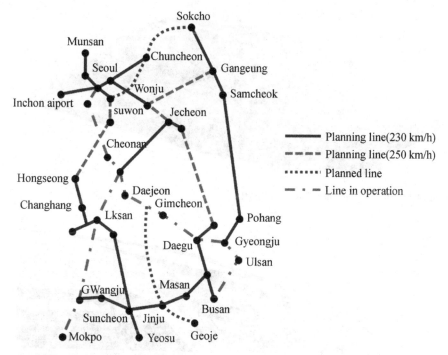

Fig. 4.25 Korea's HSR operating lines

HSR train developed by China's independent innovation realized operation and created a record of the world's highest operating speed of 486.1 km/h.

On June 26, 2017, the "Renaissance" Electrical Multiple Unit (EMU), which is one kind of WHSR trains developed by China independently, was launched in Beijing and Shanghai. The "Renaissance" EMU is one Chinese standard EMU with completely independent intellectual property rights, with a test speed of 400 km/h and above. The EMU is equipped with a powerful safety monitoring system with more than 2500 monitoring points, which are deployed throughout the vehicle to achieve comprehensive and real-time safety monitoring of the state of the running section, bearing temperature, cooling system temperature, braking system status, and passenger compartment environment. The "Renaissance" Chinese Standard EMU also adds a collision energy-absorbing device to improve the passive protection capability. It will take the lead in realizing the operating speed of 350 km/h on the Beijing-Shanghai HSR line. China has become the country with the highest HSR commercial operating speed in the world. The "Renaissance" EMU has two models: "CR400AF" and "CR400BF". The "A" and "B" are company identification codes, representing the manufacturer (Figs. 4.26, 4.27).

(1) Operation condition of HSR. In current, China's high-speed rail line network is mainly distributed in the southeast abdomen and southeast coastal areas. Among of them, there are three vertical lines, including Jing—he Line,

Fig. 4.26 The "Renaissance" CR400AF EMU

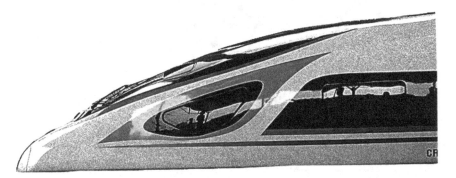

Fig. 4.27 The "Renaissance" CR400BF EMU

Shi—Wu Line and Wu—Guang Line, which connect Beijing, Tianjin, Nanjing, Shanghai, Zhengzhou, Wuhan, Changsha, Guangzhou and other capital cities from north to south. These three vertical lines are also the fast passenger transportation channels of the Yangtze River Delta Economic Circle, the Bohai Economic Circle and the Middle Reaches of the Yangtze River Economic Circle. The Ningbo—Taizhou—Wenzhou Line, Fujian—Xiamen Line, Wenzhou—Fuzhou Line and the Shenzhen—Xiamen Line under construction connect the major coastal cities such as Hangzhou, Ningbo, Wenzhou, Fuzhou, Xiamen, Shantou and Guangzhou. They are the fast passenger transportation channels of the Pearl River Delta Economic Circle. The Chengdu—Wuhan Line and Hefei—Wuhan Line are connected across central China and are the main transportation lines in the East and West. Figure 4.28 shows the Chinese HSR network.

(2) Development trend of HSR. During less than 20 years, China has developed into a country with the fastest development, the most complete system technology, the strongest integration capability, the longest operating mileage, the highest operating speed and the largest construction scale of HSR in the world. These

Fig. 4.28 Chinese HSR network topological graph

achievements are inseparable from the introduction of the advanced foreign HSR technology and the improvement of existing lines.

(3)   Recent plan of China's HSR. The HSR network will form a certain scale and complete the construction of the "four vertical lines and four horizontal lines" network to meet the spring festival transportation. China's HSR completes the speed-up passenger transport network centered on Beijing, Shanghai and Guangzhou; realizes the "morning and evening arrival" of 300–500 km distance; realizes the "evening arrival" of 1200–1500 km distance; and realizes the "one-day arrival" of 2000–2500 km distance (Table 4.11).

Table 4.11 Cities covered by Beijing traffic circle

Running time (hours)	Arrive city
1.	Tianjin, Shijiazhuang, etc
2.	Zhengzhou, Jinan, Shenyang, Taiyuan, etc
3.	Nanjing, Hefei, Changchun, Dalian, etc
4.	Shanghai, Hangzhou, Wuhan, Xi'an, Harbin, etc

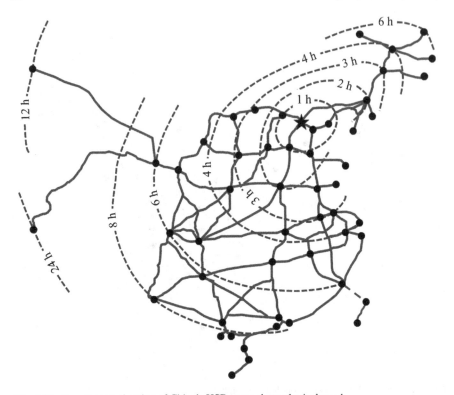

Fig. 4.29 Recent-term planning of China's HSR network topological graph

  The neighboring capital cities will form a 1–2 h traffic circle, and the provincial capital and surrounding cities will form "a half-hour" to "one-hour" traffic circle. Except Haikou, Nanning, Kunming, Urumqi, Lhasa and Taipei, the travel time from Beijing to the provincial capital cities will be within 8 h, as shown in Fig. 4.29.

(4) Medium-term planning of China's HSR. By 2020, the mileage of China's rails will reach more than 120,000 km. Among them, the new HSR lines will reach more than 38,000 km. Together with other new rails and existing speed-up lines, the China Railway Express Passenger Transport Network will reach more than 50,000 km, which connects all provincial capitals and cities of more than 500,000 people. It will cover more than 90% of the country's population, and realize the purpose of "People's walking is convenient, commodity's flow is smooth".

  According to the characteristics of the "east dense and west sparse" and the station principle of "east sparse and west dense", taking care of the west. The medium-term planning of Chinese HSR is from 2010 to 2040, using 30 years to connect the major provinces and cities across the country, forming a national network framework and forming a "five vertical lines, seven horizontal lines and eight links lines" pattern as shown in Table 4.12.

China's medium-term plan of the "five vertical" HSR lines:

The specific route of the "five verticals" in the medium-term planning of China's HSR is shown in Fig. 4.30.

Table 4.12 The "five vertical, seven horizontal and eight links" network

Index	Name	Remark
1.	Five verticals	Harbin—Shanghai Line, Beijing—Shanghai Line, Beijing—Hongkong line, Jining—Kunming Line, Xi'an—Zhanjiang Line
2.	Seven horizontals	Shenyang—Lanzhou Line, Qingdao—Yinchuan Line, Yancheng—Xining Line, Shanghai—Chengdu Line, Shanghai—Kunming Line, Shanghai—Nanning Line, Hangzhou—Guangzhou Line
3.	Eight links	Tianjin—Tangshan Line, Kaifeng—Hekou Line, Nanjing—Nantong Line, Nanjing—Ningbo Line, Jinhua—Wenzhou Line, Wuhan—Fuzhou Line, Nanping—Xiamen Line, Hengyang—Nanning Line

Fig. 4.30 China's medium-term plan of the "five vertical" HSR line topological graph

①   Harbin—Shanghai Line.

Harbin—Fuyu—Changchun—Sipingnan—Shenyang—Yingkou—Dalian—Yantai—Qingdao—Rizhao—Lianyungang (Haizhou)—Yancheng—Nantong—Shanghai. According to the above nodes, only 14 parking stations are set up, and the stations are directly connected.

②   Beijing—Hongkong Line.

Beijing—Baoding—Shijiazhuang—Handanbei—Anyangnan—Zhengzhou—Luohe—Xinyangbei—Wuhan—Yueyang—Changshanan—Hengyang—Bingzhou—Shaoguan—Guangzhou—Shenzhen—Kowloon. According to the above nodes, only 17 parking stations are set up, and the stations are directly connected.

③   Jining—Kunming Line.

Jining—Datong—Shuozhou—Xinzhoubei—Taiyuannan—Jiexiu—Linfen—Hancheng—Xi'an—Foping—Hanzhong—Ningqiang—Guangyuan—Mianyang—Chengdu—Leshan—Mianning—Xichang—Panzhihua—Kunming. According to the above nodes, only 20 parking stations are set up, and the stations are directly connected.

④   Xi'an—Zhanzhou Line.

Xi'an—Ankang—Wanyuan—Dazhou—Huaying—Chongqing—Zunyi—Guiyang—Duyun—Dushan—Nandan—Chihexi—Mashanbei—Nanning—Qinzhou—Beihai—Zhanjiang. According to the above nodes, only 17 parking stations are set up, and the stations are directly connected.

⑤   Beijing—Shanghai Line.

Beijing—Tianjin—Cangzhou—Dezhou—Jinanxi—Jining—Xuzhou—Bengbu—Nanjing—Wuxi—Shanghai—Pudong airport.

China's medium-term plan of the "seven horizontals" HSR line:
   The specific line of "seven horizontals" in the medium-term planning of Chinese HSR is shown in Fig. 4.31.

①   Shenyang—Lanzhou Line.

Shenyang—Panjin—Jinzhou—Qinhuangdao—Tangshan—Beijing—Zhangjiakou—Jining—Hohhot—Baotou—Hangjin—Wuhai—Shizuishan—Yinchuan—Qingtongxia—Zhongwei—Baiyin—Lanzhou. According to the above nodes, only 20 parking stations are set up, and the stations are directly connected.

②   Qingdao-Yinchuan Line.

Qingdao—Weifang—Zibo—Jinanxi—Wucheng—Hengshui—Shijiazhuang—Yangquan—Taiyuannan—Lvliang (Lishi)—Suide—Jinbian—Otog—Yinchuan. According to the above nodes, only 14 parking stations are set up, and the stations are directly connected.

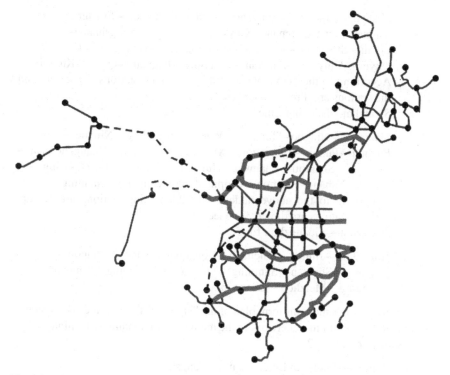

Fig. 4.31 China's medium-term plan of the "seven horizontal" HSR line topological graph

③ Yancheng—Xining Line.

Yancheng—Huai'an—Suqian—Xuzhouxi—Shangqiu—Kaifengdong—Zhengzhou—Luoyang—Sanmenxia—Huayin—Xi'an—Baoji—Tianshui—Dingxi—Lanzhou—Honggu—Xining. According to the above nodes, only 17 parking stations are set up, and the stations are directly connected.

④ Shanghai—Chengdu Line.

Shanghai—Nanjing—Hefei—Lu'an—Macheng—Wuhan—Qianjiang—Jingzhou—Yichang—Shuibuya (or Wufeng)—Enshi—Qianjiang—Peilingxi—Chongqing—Suining—Chengdu. According to the above nodes, only 15 parking stations are set up, and the stations are directly connected. The line goes southeast, passing Suyang–Huzhou–Hangzhou–Shaoxing–Ningbo, and eastward can be along north of the Yangtze river, passing Yangzhou, Taizhou to Nantong.

⑤ Shanghai—Kunming Line.

Shanghai—Jiaxing—Hangzhou—Jinhua—Quzhou—Shangrao—
Yingtan—Nanchangnan—Xinyu—Pingxiang—Changshanan—
Loudi—Shaoyang—Dongkoubei—Huaihua—Yuping—Kaili—
Duyun—Guiyang—Anshun—Guanling—Panxian—Qujin—Kunming.
According to the above nodes, only 24 parking stations are set up, and
the stations are directly connected.

⑥   Shanghai—Nanning Line.

Shanghai—Ningbo—Taizhou—Wenzhou—Fuding—Ningde—
Fuzhou—Putian—Quanzhou—Xiamen   (Tong'an)—Zhangzhounan—
Yunxiao—Shantou—Shanwei—Huizhou—Guangzhou—Zhaoqing—
Yunfu—Yunan—Wuzhou—Guipingdong—Guigang—Nanning.
According to the above nodes, only 23 parking stations are set up,
and the stations are directly connected.

⑦   Hangzhou—Guangzhou Line.

Hangzhou—Jinhua—Suichang—Longquan—Songxi—Jianou—
Nanping—Shaxian—Sanming—Yong'an—Zhangping—Longyan—
Yongding—Meizhou—Guangzhou.

Thirdly, China's medium-term plan of the "eight links" HSR line. The specific
line of the "eight links" network in the medium-term planning of China's HSR
is shown in Fig. 4.32.

①   Tianjin—Tangshan Line: Tianjin—Tangshan.
②   Kaifeng—Hekou Line: Kaifengdong—Heze—Dongping—Jinanxi—
     Binzhou—Dongyingbei—Hekou.
③   Nanjing—Nantong Line: Nanjing—Yangzhou—Taizhou—Nantong.
④   Nanjing—Ningbo Line: Nanjing—Shuyang—Huzhou—Shaoxing—
     Ningbo.
⑤   Jinhua—Wenzhou Line: Jinhua—Lishui—Wenzhou.
⑥   Wuhan—Fuzhou Line: Wuhan—Huangshixi—Wuxue (Jiangnan)—
     Jiujiangxian—De'an—Nanchangnan—Fuzhou (Jiangxi)—Shaowu—
     Nanping—Fuzhou.
⑦   Nanping—Xiamen Line: Nanping—Sanming—Datian—Xiamen
     (Tong'an).
⑧   Hengyang—Nanning Line: Hengyang—Qidong—Yongzhou—
     Quanzhou—Liuzhou—Laibin—Binyang—Nanning.

(5)   Long-term planning of China's HSR. The long-term planning of China's HSR
      is from 2040 to 2070, and it will take another 30 years to complete at the latest
      2100. The east dense is realized, and the west is connected into a network
      (that is, the major transportation hubs in the west is connected). The major
      transportation node cities and tourist attractions are connected, so that major
      cities in the western region can access any coastal provinces and regions, and the
      WHSR Network of "eight vertical lines and eight horizontal lines" is realized
      as seen in Fig. 4.33.

Fig. 4.32 China's medium-term plan of the "eight links" HSR line topological graph

China's long-term planning of the "eight vertical" HSR lines is as shown in Fig. 4.34.

① Coastal Channel. HSR of Dalian (Dandong)—Qinhuangdao—Tianjin—Dongying—Weifang—Qingdao (Yantai)—Lianyungang—Yancheng—Nantong—Shanghai—Ningbo—Fuzhou—Xiamen   (Fangchenggang) (one of them, the line of Qingdao to Yancheng is using Qinglian line, Lianyan rail line, the section of Nantong to Shanghai in Shanghai-Nantong rail). It connects the eastern coastal areas and runs through the urban agglomerations of Beijing-Tianjin-Hebei, South central of Liaoning, Shandong Peninsula, Dongpuhai, the Yangtze River Delta, the west coast of the Straits, the Pearl River Delta and Beibu Gulf.

② Beijing—Shanghai Channel. The HSR line of Beijing—Tianjin–Jinan—Nanjing—Shanghai (Hangzhou), which includes the HSR line of Nanjing—Hangzhou and Bengbu—Hefei—Hangzhou, at the same time, it connects to the HSR line of Beijing—Tianjin—Dongying—Weifang—Linyi—Huaian—Yangzhou—Nantong—Shanghai, also connects to North China and East China, and penetrates Beijing—Tianjin—Hebei, Yangtze River Delta and other urban agglomerations.

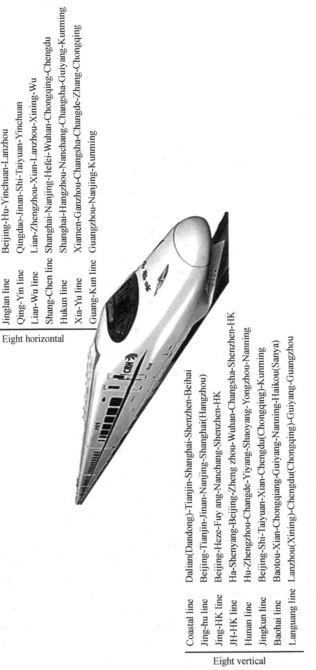

Eight horizontal	
Sui-Man line	Sui-Ha-Qi-Hai-Man
Jinglan line	Beijing-Hu-Yinchuan-Lanzhou
Qing-Yin line	Qingdao-Jinan-Shi-Taiyuan-Yinchuan
Lian-Wu line	Lian-Zhengzhou-Xian-Lanzhou-Xining-Wu
Shang-Chen line	Shanghai-Nanjing-Hefei-Wuhan-Chongqing-Chengdu
Hukun line	Shanghai-Hangzhou-Nanchang-Changsha-Guiyang-Kunming
Xia-Yu line	Xiamen-Ganzhou-Changsha-Changde-Zhang-Chongqing
Guang-Kun line	Guangzhou-Nanjing-Kunming

Eight vertical	
Coastal line	Dalian(Dandong)-Tianjin-Shanghai-Shenzhen-Beihai
Jing-hu line	Beijing-Tianjin-Jinan-Nanjing-Shanghai(Hangzhou)
Jing-HK line	Beijing-Heze-Fuy ang-Nanchang-Shenzhen-HK
JH-HK line	Ha-Shenyang-Beijing-Zheng zhou-Wuhan-Changsha-Shenzhen-HK
Hunan line	Hu-Zhengzhou-Changde-Yiyang-Shaoyang-Yongzhou-Nanning
Jingkun line	Beijing-Shi-Taiyuan-Xian-Chengdu(Chongqing)-Kunming
Baohai line	Baotou-Xian-Chongqiang-Guiyang-Nanning-Haikou(Sanya)
Languang line	Lanzhou(Xining)-Chengdu(Chongqing)-Guiyang-Guangzhou

Fig. 4.33 "Eight horizontal and eight vertical" network of China's HSR

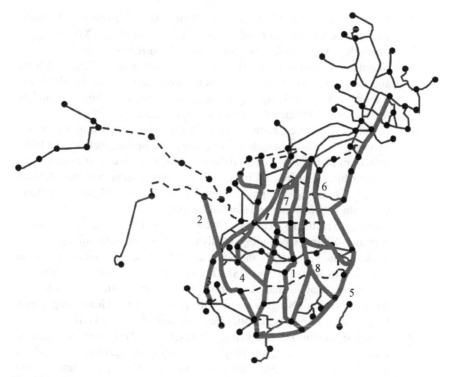

Fig. 4.34 "Eight vertical" HSR line topological graph

③ Beijing—Hongkong (Taiwan) Channel. The HSR line of Beijing—Heng-shui—Heze—Shangqiu—Fuyang—Hefei (Huanggang)—Jiujiang—Nanchang—Ganzhou—Shenzhen—Hongkong (Kowloon); Another line is: Hefei—Fuzhou—Taipei, including the railway of Nanchang—Fuzhou (Putian). It connects North China, Central China, East China and South China, and runs through Beijing—Tianjin—Hebei, the middle reaches of the Yangtze River, the west coast of the Straits, and the Pearl River Delta.

④ Beijing—Harbin and Beijing—Hongkong—Macao Channel. The HSR line of Harbin—Changchun—Shenyang—Beijing—Shiji-azhuang—Zhengzhou—Wuhan—Changsha—Guangzhou—Shen-zhen—Hongkong, includes the HSR line of Guangzhou—Zhuhai—Macao. It connects Northeast China, North China, Central China, South China, Hong Kong and Macao, and integrates urban groups such as Harbin—Changchun, South Central Liaoning, Beijing—Tianjin—Hebei, Central plains of China, the middle reaches of the Yangtze River and the Pearl River Delta.

⑤ Hohhot—Nannjing Channel. The HSR line of Hohhot—Datong—
Taiyuan—Changzhi—Jincheng—Jiaozuo—Zhengzhou—Xiangyang—
Changde—Yiyang—Loudi—Shaoyang—Yongzhou—Guilin—
Nanning. It connects North China, Central plains, Central China,
and South China, and runs through urban areas such as Hohhot—
Baotou—Ordos—Chongqing, Central Shanxi, Central Plains, the middle
reaches of the Yangtze River, and the Beibu Gulf.

⑥ Beijing—Kunming Channel. The HSR line of Beijing—Shijiazhuang—
Taiyuan—Xi'an—Chengdu (Chongqing)—Kunming, includes the high-
speed rail of Beijing—Zhangjiakou—Datong—Taiyuan. It connects the
North China, Northwest China, and Southwest China, and runs through
the city clusters of Beijing-Tianjin–Hebei, Taiyuan, Guanzhong Plain,
Chengdu—Chongqing and Central Yunnan.

⑦ Baotou (Yinchuan)—Haikou Channel. The HSR line of Baotou—
Yan'an—Xi'an—Chongqing—Guiyang—Nanning—Zhanjiang—
Haikou (Sanya), which includes the HSR line of Yinchuan—Xi'an
and Hainan Round Island HSR. It connects the Northwest, Southwest,
and South China, and runs through the city clusters of Hohhot—
Baotou—Ordos, Ningxia along the Yellow River, Guanzhong Plain,
Chengdu—Chongqing, Central Guizhou, and the Beibu Gulf.

⑧ Lanzhou (Xining)—Guangzhou Channel. The HSR line of Lanzhou
(Xining)—Chengdu(Chongqing)—Guiyang—Guangzhou,   which
connects the Northwest, Southwest, and South China, and runs through
the city clusters such as Lanzhou—Xining, Chengdu—Chongqing,
Central Guizhou, and the Pearl River Delta.

China's long-term planning of the "eight horizontal" HSR lines is as shown in
Fig. 4.35.

① Suifenhe—Manzhouli Channel. The HSR line of Suifenhe—Mudan-
jiang—Harbin—Qiqihar—Hailar—Manzhouli, which connects the area
of Heilongjiang and East Inner Mongolia.

② Beijing—Lanzhou Channel. The HSR line of Beijing—Hohhot—
Yinchuan—Lanzhou, which connects to the North, Northwest of China,
runs through the city clusters of Beijing—Tianjin—Hebei, Hohhot—
Baotou—Ordos, Ningxia along the Yellow River, Lanzhou—Xining.

③ Qingdao—Yinchuan Channel. The HSR line of Qingdao—Jinan—Shijiazhuang—Taiyuan—Yinchuan (Among of them, the section of Suide to
Yinchuan is using the rail line of Taiyuan—Zhongwei), which connects
the East, North, Northwest of China, runs through the city clusters of
Shandong Peninsula, Beijing—Tianjin—Hebei, Taiyuan, Ningxia along
the Yellow River.

④ Along the Yangtze River Channel. The HSR line of Shanghai—
Nanjing—Hefei—Wuhan—Chongqing—Chengdu, which includes HSR
line of Nanjing—Anqing—Jiujiang—Wuhan—Yichang—Chongqing,
Wanzhou—Dazhou—Suining—Chengdu (Among of them, the section

Fig. 4.35 "Eight horizontal" HSR line topological graph

of Chengdu to Suining is using the railway of Dacheng—Chengdu). It includes the East, Central and Southwest of China, runs through the city clusters of Yangtze River Delta, Middle reaches of the Yangtze River and Chengdu—Chongqing.

⑤ Shanghai–Kunming Channel. The HSR line of Shanghai—Hangzhou—Nanchang—Changsha—Guiyang—Kunming, which connects the East, Central, Southwest of China, runs through the city clusters of the Yangtze River Delta, Middle reaches of the Yangtze River and Central Guizhou, Central Yunnan.

⑥ Xiamen—Chongqing Channel. The HSR line of Xiamen—Longyan—Ganzhou—Changsha—Changde—Zhangjiajie—Qianjiang—Chongqing (Among of them, the section of Xiamen to Ganzhou is using the rail line of Longyan—Xiamen, Ganzhou—Longyan, and the section of Changde to Qianjiang is using the rail line of Qianjiang—Zhangjiajie—Changde). It connects the west coast of the Taiwan Straits, South Central, Southwest of China, runs through the city clusters of the west coast of the Taiwan Straits, Middle reaches of the Yangtze River and Chengdu–Chongqing.

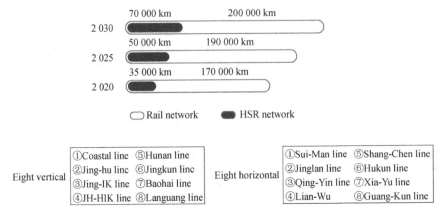

Fig. 4.36 China's WHSR lines (picture from the network)

⑦ Guangzhou—Kunming Channel. The HSR line of Guangzhou—Naning—Kunming, which connects the South, Southwest of China, runs through the city clusters of the Pearl River Delta, Beibu Gulf, and the Central Yunnan.

⑧ Luqiao Channel. The HSR line of Lianyungang—Xuzhou—Zhengzhou—Xi'an—Lanzhou—Xining—Urumqi, which connects the East, Central, Northwest of China, runs through the city clusters of east Longhai, Central plain, Guanzhong plain of China, Lanzhou—Xining and the northern slope of Tianshan Mountain. China's railway is developing continuously, and the construction of China's "eight vertical and eight horizontal" HSR network is gradually improved. The situation of China's WHSR lines is shown in Fig. 4.36.

4.2.3 Development Trend of WHSR in America

Currently, there is no HSR in America, of which the United States may be the first country in North America to have HSR, and Brazil may be the first country in South America to have HSR.

(1) HSR in the United States. The United States was once the king of the rails in the world. The US rail freight is still in the leading position in the world, but the development of HSR in the United States is particularly slow: it proposed the HSR construction plan in the 1960s, but the Western California Interstate HSR is expected to be completed as early as 2020. There are several reasons for the slow development of HSR in the United States: low density population distribution, highly developed road network, the high penetration rate of automobiles and developed aviation industries, etc.

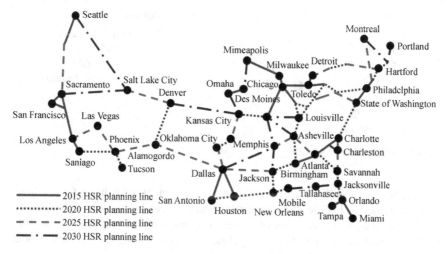

Fig. 4.37 HSR planning map of The Unite States

The United States hopes to create 11 HSR channels, three in the densely populated northeastern region, and then extends to Florida, the Northern coast of Mexico Gulf, the Midwest, Texas, the Pacific Northwest and California. The United States will develop one HSR network of 27,000 km in the next 20–30 years, as shown in Fig. 4.37.

(2) HSR in Brazil. The Brazilian government has placed the development of rail transport at a priority over the investment in large infrastructure projects and has developed a medium-term plan for the rail network. The main content of the plan is to improve the rail network structure, including rail safety system, rail speed increase, suburban transportation, public rail intersect bridge, housing demolition and residential resettlement along the rail, etc. It is expected to invest 1 billion Real in 5 years. At present, only one HSR line in Brazil is under planning, as shown in Fig. 4.31. It is expected to depart from Campinas and pass through 11 cities in the middle of São Paulo, Jacarere, Taubate and Cruzema, and finally arrive in Rio de Janeiro with a total mileage of 511 km and a planned speed of 285 km/h. It will shorten 1.5 h from São Paulo to Rio de Janeiro. Figure 4.38 shows the Brazil's first HSR line.

4.3 Development Vision of WHSR

The development of HSR has a major impact on national and regional development strategies. At present, many countries in the world have begun to build transnational HSR, as soon as possible to achieve rapid road access between countries and regions, eliminate the impact of geographical restrictions between countries, and accelerate exchanges and cooperation of economy and resources between regions and regions.

Fig. 4.38 Brazil's first HSR line

Considering the overall situation of national development, HSR has a far-reaching strategic influence. For the strategic, national security can be guaranteed. Especially considering the overall situation of global development, the development of HSR has a profound impact on the world political economy. HSR can promote world integration and realize "global village".

As a safe and reliable, fast and comfortable, large-capacity, low-carbon and environmentally-friendly transportation mode, HSR has become an important trend in the development of the world transportation industry. At present, 9 countries own HSR in Europe are planning to invest 200 billion US dollars for extending the HSR with a total mileage of 7000 km to 16,000 km. Japan has already launched the construction of the magnetic levitation central Shinkansen, which will connect Tokyo and Osaka. The train running on this line will become the fastest in the world at speed of 500 km/h. The United States, the traditional country that based on road transport, now is starting to develope HSR. By 2030, it will cover 80% of Americans. The development of China's HSR is from scratch, and China has also changed from a seeker to a front-runner. The scale and speed of construction are at the forefront of the world. By 2020, 90% of Chinese people can travel by HSR. Therefore, in 2020, the world will enter not only the "era of high-speed rail", but also the era of "global village".

4.4 Summary

The punctuality, speed and comfort of HSR are favored by tourists from all over the world and the HSR has become the preferred mode of transportation for travel. Comparing with other modes of transportation, HSR has strong conveying capacity, fast speed, good safety, highly punctuality rate, low energy consumption, less impact on the environment, less land occupation, comfort and convenience, considerable economic benefits and good social benefits. With its unique technical advantages to adapt the new demand of modern social and economic development, the HSR has become an inevitable choice for the development of countries around the world. The development and operation practice of China's HSR shows that it has great development space and potential in China. China should make full use of its latecomer advantage to realize the leap-forward development of China's HSR.

Chapter 5
Magnetic High-Speed Rail (MHSR)

Affected by the air resistance and the track friction, the threshold value of operating speed of WHSR is 400 km/h. If the threshold value exceeds 400 km/h, the WHSR is easy to derail and cause traffic accidents. Then, how can we make the HSR run faster? One idea is to let the HSR train float, so there is no friction between the HSR train and the track. Based on the above ideas, the specialists began to study the magnetic levitation technology. Figure 5.1 shows the Japan's MHSR train.

The study of the magnetic levitation technology was originated in Germany. In 1922, German engineer Hermann Kemper proposed the electromagnetic floating principle. He believed that since the maximum resistance of the train came from the friction between the train wheels and the wheel track, would it ran faster if the train could levitate above the track? In 1934, Herman received the world's first patent on magnetic technology. After 1970, with the increasing economic strength of the industrialized countries in the world, developed country such as Germany, Japan, the United States, Canada, France and the United Kingdom have begun to plan the development of Magnetic Transport System in order to enhance their transport capacity and meet the needs of economic development. The United States and the Soviet Union abandoned the research plan in the 1970s and 1980s respectively. At present, only Germany, Japan, China, and South Korea continue to conduct research on magnetic levitation systems, and they have all made remarkable progress.

One HSR system that operates using the magnetic levitation technology is called the Magnetic High-speed Rail (MHSR). Unlike the general wheel-rail adhesive rail, the MHSR train has no wheels, and the vehicle is suspended on the rail surface of the track by means of no contact magnetic technology. Therefore, the MHSR train is of one train that is driven by magnetic levitation forces (magnetic suction and repulsive forces). MHSR is suspended in the air by the magnetic force of the orbit, and there is no contact with the ground during operation, so it is only affected by the resistance of the air. The maximum speed of the MHSR train can reach more than 500 km/h, which is faster than the WHSR train. World's first operational magnetic levitation line is the Shanghai Pudong Airport Line in China as shown in Fig. 5.2.

© Southwest Jiaotong University Press 2023
Q. Hu and S. Qu, *A Brief History of High-Speed Rail*,
https://doi.org/10.1007/978-981-19-3635-7_5

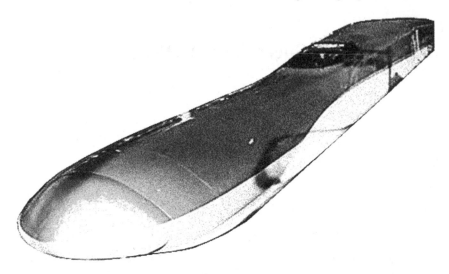

Fig. 5.1 MHSR train

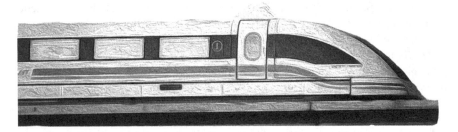

Fig. 5.2 Shanghai MHSR train

Japan's MHSR train adopts the superconducting magnetic levitation technology. In November 1982, the manned test of the Japanese MHSR train was successful. In 1997, the Yamanashi magnetic suspension test line of 18.4 km was successfully built and Japan began the running tests. On December 2, 2003, three MLX01 MHSR trains in Japan created a world record of 581 km/h in Yamanashi. However, the construction of the magnetic levitation lines has not been approved due to high cost. On the morning of April 21, 2015, the vehicle-driving test of superconducting MHSR train was carried out in Yamanashi, setting a record high of 603 km/h and continuously refreshing the highest speed record of land-based manned vehicles (Fig. 5.3).

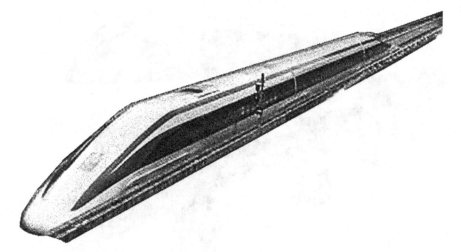

Fig. 5.3 Japan's MHSR train

5.1   Basic Principles of the Magnetic Levitation Technology

The basic principle of the magnetic levitation is very simple. It uses the electromagnetic floating principle of "same-magnet repelling and opposite-magnet attraction" to magnetize the gravitational force against the gravity, so that the vehicle can be levitated, and finally guided by electromagnetic force to push the train forward. From a technical point of view, magnetic levitation mainly includes three technologies: no contact support, guiding technology and driving technology. Due to technical and technological constraints, it was not until the 1960s that developed countries began to conduct large-scale research on the magnetic transportation. Germany and Japan were the ones that invested more energy and achieved more outstanding achievements. Figure 5.4 shows the architecture of MHSR.

(1) Superconductor. Scientists have discovered that many metals and alloys completely lost electrical resistance at specific low temperatures. Conductors with such property are called cryogenic conductor. In 1911, Kadulin-Onnes of the University of Leiden in the Netherlands unexpectedly discovered that when mercury was cooled to subzero 268.98 °C, the resistance of mercury suddenly disappeared. Kamenlin-Annes called it a superconducting state. Because of this discovery, he won the Nobel Prize in 1913 (Fig. 5.5).

(2) Meissner Effect. In 1933, the German physicists Meissner and Ossenfeld measured the magnetic field distribution of the tin single crystal superconductor. When the metal was cooled into the superconducting state in a small magnetic field, the magnetic lines of the body were discharged and the magnetic lines could not pass its body. That is to say when the superconductor is in superconducting state, the magnetic field in the body is always equal to zero. This effect is known as the "Meissner effect" (Fig. 5.6).

Fig. 5.4 Architecture of MHSR

Fig. 5.5 Superconducting
state

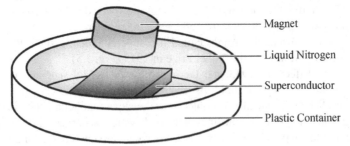

Magnet

Liquid Nitrogen

Superconductor

Plastic Container

Fig. 5.6 Meissner effect

(3)   Principle of repulsive force. The superconductor "does not allow" any magnetic field inside it. If there is an external magnetic field passing through the super-conductor, the superconductor will inevitably generate a magnetic field oppo-site to it, ensuring that the internal magnetic field strength is zero, which forms a repulsive force. When a magnet is placed directly below a superconductor and the magnetic induction lines are passed vertically through the supercon-ductor, the superconductor will obtain vertical upward buoyancy. When this force is exactly equal to the gravity of the superconductor, the superconductor can be suspended in the air.

The repulsive force gradually increases as the relative distance decreases, and it can overcome the gravity of the superconductor and make the MHSR train suspend at a certain height above the permanent magnet. When the super-conductor moves away from the permanent magnet, it produces a negative flux density in the superconductor and induces a reverse critical current, which generates suction on the permanent magnet. This suction can overcome the gravity of the superconductor and make the MHSR train hang upside down in a certain position under the permanent magnet (Fig. 5.7).

(4)   Principle of MHSR train. At present, Japan and Germany are in the leading position in magnet field. Japan is Electrodynamic Suspension (EDS), whose highest test speed reached 603 km/h. Germany is Electromagnetic Suspension (EMS), whose highest test speed is 505 km/h. The Shanghai MHSR line uses Germanic technology and operates at a speed of 430 km/h. Figure 5.8 shows the architecture of MHSR train.

The speed of the MHSR train reaches 500 km/h, which is absolutely impos-sible for traditional trains. If the superconducting magnet is installed in the

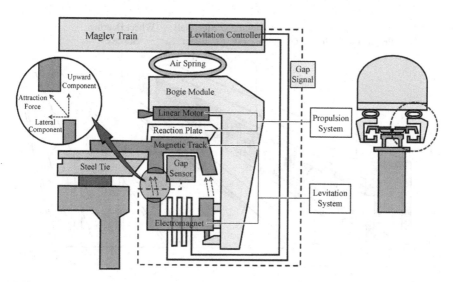

Fig. 5.7 Suspension system (picture from the network)

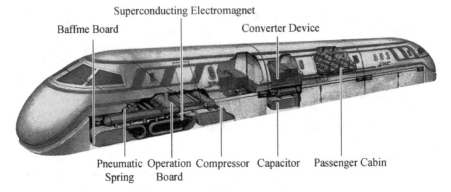

Fig. 5.8 Architecture of MHSR train (picture from the network)

train and the aluminum ring is laid on the ground track, the relative movement between them will generate the induced current occur in the aluminum ring, thus producing the magnetic repulsion. The magnetic repulsion will lift the train about 10 cm above the ground and the train can levitate on the ground and move at a high speed. Figure 5.9 shows the architecture of magnetic orbit.

MHSR trains generally have low temperature superconductors (-4.2 K liquid helium is rare and expensive) and high temperature superconductor (-77 K liquid nitrogen is more and cheaper), among them the best temperature is -273.15 °C. The MHSR train can resist the gravity of the earth and suspend on the track. According to the working principle, it can be divided into normal conductor electromagnetic attraction levitation and superconducting

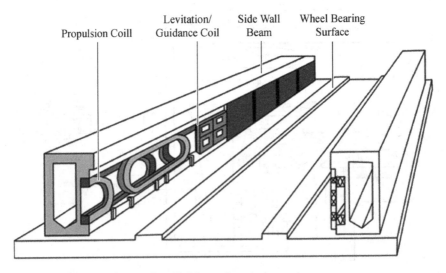

Fig. 5.9 Architecture of magnetic orbit (picture from the internet)

electromagnetic repulsive levitation. "MHSR train" is characterized by fast, low consumption, environmental protection and safety. The train covers the track and runs on the track, avoiding the danger of derailment, so the safety is extremely high. The operating power of HSR train comes from the electromagnetic flow fixed on both sides of the track, and the electromagnetic flow intensity in the same area is the same. It is impossible to have several trains' speeds or opposite movements, thus eliminating the possibility of rear-end or collision. However, the total cost of constructing a MHSR line is very high, which is equivalent to the construction cost of three WHSR lines.

(5)   Principle of Electromagnetic Suspension (EMS). EMS is of an electromagnetic active control suspension, which generates the electromagnetic attraction from the constant current flow on the vehicle. It attracts the magnets below the track, so that the train can be suspended and then driven forward by the linear motor.

The track in Fig. 5.10 is a "T" shaped table. The lower part of the train wraps the two sides of the "T" shaped track, which is to supply current to the suspended electromagnet coil placed under the guide rail to generate an electromagnetic field, so that the suspended electromagnet coil can interact with the ferromagnetic guide rail on the track, and use the electromagnetic attraction between them to make the train suspend to a certain height. However, due to the nonlinear relationship between the electromagnetic attraction and the size of the air gap: a decrease in the air gap increases the electromagnetic attraction, causing the air gap to further decrease, while an increase in the air gap causes the electromagnetic attraction to decrease, resulting in a further increase in the air gap. The German TR-type magnetic train is a typical representative of electromagnetic suspension (Table 5.1).

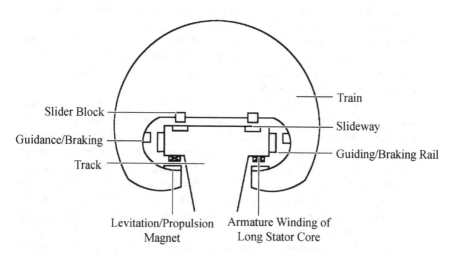

Fig. 5.10 Germanic MHSR train, drive and guiding mechanism

Table 5.1 Characteristics of electromagnetic suspension

Name	Characteristics
Advantage	Simple technology
	Speed 400–500 km/h
Disadvantage	The system is inherently unstable and produces less electromagnetic attraction
	The gap between the train and the track is generally 8–10 mm
	Accurate and fast feedback control is required to ensure reliable and stable suspension of trains

(6)   Principle of Electrodynamic Suspension (EDS). EDS uses the principle of mutual exclusion between the same magnetic poles to realize vehicle suspension. Since the root of the gravitational resistance is that the magnetic field of the induced current and the magnetic field of the superconducting coil repel each other to generate a repulsive force, the higher the speed of the train, the greater the repulsive force. When the speed exceeds a certain value, the train is suspended from the surface of the track. The principle is to install a superconducting coil or a permanent magnet on the body of the MHSR train, and a "8"shaped coil arranged in a regular arrangement is distributed on the track. When the train runs at a certain speed, the strong magnetic field is generated by the superconducting coil. It will generate current in the "8" shaped coil of the track, and then the induced current generates a strong electromagnetic field, forming a repulsive magnetic field in the lower half of the "8" shaped coil and an attracting magnetic field in the upper half of the "8" shaped coil, so that the train can be suspended. The EDS is not a train that wraps track, but a track wraps the train. It uses a vehicle-mounted superconducting magnet to generate mutual repulsive force with the induced magnetic field of the track during the movement, and is to suspend in the track. The train is operating in a "U" shaped groove. Japan Tokai Railway Company's MLU train with the speed of 603 km/h in the Yamanashi Line is the representative of the EDS train, as shown in Fig. 5.11 (Table 5.2).

(7)   Principle and classification of the magnetic traffic technology. Magnetic traffic, just as its name implies, is a mode of transportation that relies on magnetic force to make the MHSR train suspend and operate. Electromagnetic suspension is to generate an electric field by energizing and exciting a suspended electromagnet that placed on the underside of a track. The magnet and the ferromagnetic member on the track are attracted by each other, and the train is sucked up and suspended on the track. The suspension gap between the magnet and the ferromagnetic track is generally 8–12 mm. The train guarantees a stable air gap of suspension by controlling the excitation current of the suspended magnet, and the train is towed by a linear motor. The working mode of the electric suspension is that when the train moves, the moving magnetic field of the vehicle magnet generates an induced current in the suspension coil installed on the line, and the two interact to generate an upward magnetic force to suspend

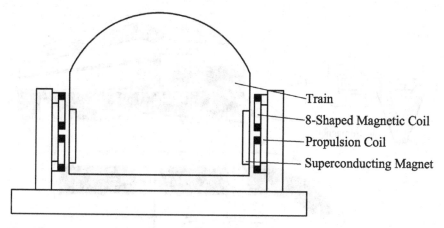

Fig. 5.11 The technology of EMS

Table 5.2 Characteristics of electrodynamic suspension

Name	Characteristics
Advantage	Large levitation force, fast train running speed, most is high-speed magnetic levitation column
	Realize speed above 500 km/h
	Well adapt to the natural conditions of mountainous terrain and frequent earthquakes
Disadvantage	Suspension gap: about 100 mm
	Technology is complex, need to shield the divergent electromagnetic field

the train at a certain height on the road surface (generally 100–150 mm). The train is operated by linear motor traction. Comparing with electromagnetic suspension, the electric magnetic system cannot be suspended at rest. It must reach a certain speed before levitation, which is about 150 km/h (Fig. 5.12).

Traction motors of MHSR trains are linear motors. According to the motor form, it can be divided into two types: namely, the long-stator linear synchronous motor and the short-stator linear induction motor. When a long-stator linear synchronous motor is used, the stator of the motor is laid along the entire line and the rotor of the motor is mounted on the vehicle, which is suitable for the traction of MHSR train, such as the Shanghai demonstrate MHSR line, the Germanic TR EDS train and Japan's MLX, LO and other series of MHSR trains. When a short-stator linear induction motor is used, the stator of the motor is mounted on the vehicle and the rotor is on the track, which is suitable for low speed magnetic trains, such as Changsha MHSR line, Beijing MHSR line and Japan's Aichi MHSR line. Therefore, the technology of MHSR traffic can be divided into high-speed magnetic technology (maximum speed is about 400–500 km/h) and medium–low speed magnetic technology (maximum speed is

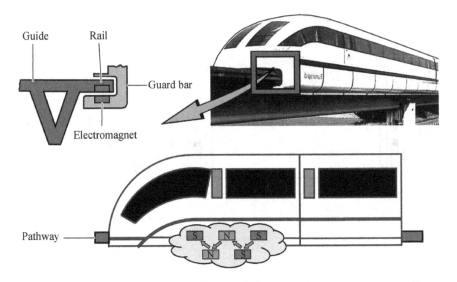

Fig. 5.12 Operation diagram for MHSR train (picture from the network)

generally 80–100 km/h). Among them, the magnetic suspension at a development speed of 200 km/h is called a medium-speed magnetic technology (Fig. 5.13).

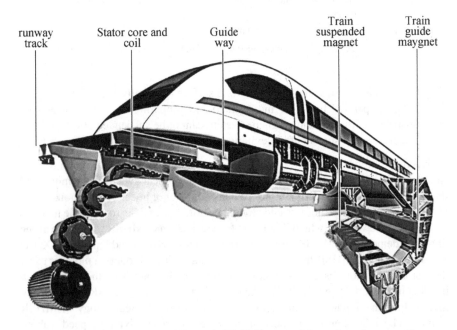

Fig. 5.13 Structure and principle of Germanic TR EMS system (picture from the network)

Japan's superconducting MHSR train uses a low-temperature (absolute temperature is 4.2 K) superconducting coil mounted on a vehicle to generate a strong magnetic field, which is called superconducting magnetic traffic. In addition, other magnetic traffic technologies currently use ordinary conductors to electrify to generate electromagnetic buoyancy and guiding force, so it is called the constant-conducting magnetic traffic.

5.2 Main Characteristics of Magnetic Traffic

As a kind of public transportation mode, the operation principle of magnetic traffic system is similar to the rail (including HSR) and urban rail transit, and the subsystem structure is basically the same. In generally, it can be divided into four subsystems: operation control system, vehicle system, traction power supply system, and line track system. In addition to the characteristics of non-contact "suspension" operation, the magnetic traffic is a highly integrated automatic control system, which is showed in Fig. 5.14. Taking the MHSR technology as an example, and its main features are:

(1)   Safety. MHSR trains ensure safety through a number of safety design techniques. The control and safety protection technology of the train operation ensures that any faults that hinder normal operation will lead to safe stop; no rear-end collision and collision will occur; even in extreme fault conditions, suspension and braking can be guaranteed.

   ①   No derailment. The system structure of vehicles running around the line avoids the possibility of derailment.

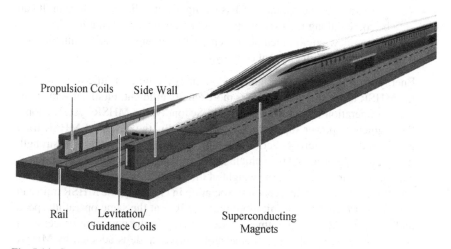

Fig. 5.14 Construction of the guide rail of MHSR

② High fire resistance. The vehicle is designed according to the fire protection standard of the civil aircraft, and the fire isolation time of the isolation door between adjacent cars is more than 10 min.

③ High stability. The vehicle adopts anti-collision design to ensure that the suspension and guiding stability of the train will not be jeopardized after the collision of possible range, and the deformation of the train collision part will not crush the personnel and passengers in the vehicle.

④ High shock resistance. Due to the high acceleration and deceleration performance of the system, the train can be parked with a shorter braking time after receiving the seismic monitoring signal.

(2) Environmental protection. The operation of MHSR system imposes less burden on the environment than other comparable traffic systems. Moreover, the MHSR line has a small enclosed area, small required surface area, and low energy consumption ratio, therefore the carbon dioxide emissions and acoustic emission are also superior.

① Low noise. At the same speed, the noise of MHSR train is significantly lower than other land vehicles.

② Suitable Radiation. According to the overall and comprehensive measurement conducted by the US and Chinese authorities on the Shanghai demonstrate MHSR line and the comparison with the international standards and national standards of electromagnetic radiation, the results show that: the electromagnetic radiation of the MHSR system is equivalent to the WHSR system and the subway system, which is far below the current standard limits and will not adversely affect the health of the public and professional personnel.

③ Land conservation. Because of its relatively high climbing ability and small turning radius, the line has strong adaptability to the terrain. It can be routed along the existing traffic corridor for occupying less land, not adding new environmental burden, and reducing the noise influence on the public and the natural ecology.

(3) Energy saving. The rail is recognized as an energy-efficient traffic system, while the MHSR transportation system is more energy efficient. Firstly, it is the non-contact operation mode of MHSR train; Secondly, the MHSR system adopts the segmented power supply technology; Thirdly, the design of MHSR train greatly reduces the cross-sectional area of the train; Finally, the high-strength lightweight material and the vehicle body structure on the magnetic train uses similar measures to reduce the weight of the aircraft. According to research by the Rhine Technical Supervision Association in Germany, the MHSR train can save 20–40% energy per unit more than the ICE at the same operating speed and seat size. At the same time, the feed energy in the operation process can be recycled; the synchronous linear motor drive mode is adopted by MHSR train, and its power supply efficiency is also higher.

(4) Reliability. MHSR traffic is a highly automated and informative active control system. Train's operation, control and maintenance are automated. Based on diagnostic technology, the informatization of operation, maintenance and management are achieved. Due to the automation and information features of the system technology, its operational reliability is relatively higher. Due to its own technical characteristics, the MHSR line balances the centrifugal force by setting the cross slope on the track surface. The maximum cross slope of the section can be $\leq 12°$ ($\alpha \leq 16°$ in special cases), the turning radius of the line is small, and it has strong climbing ability (Longitudinal slope $\leq 10\%$). The above characteristics make the MHSR line more adaptable, the surrounding facilities and terrain conditions have less restriction on route selection. In other aspects such as comfort and national economic benefits, the MHSR system also has certain advantages. The measured comfort level of the Shanghai demonstration line has reached the highest level of ISO comfort standards, and so on.

(5) Applicability. Due to the difference in speed, the application range of high-speed magnetic and medium–low speed magnetic traffic technology is different. The travel speed of the MHSR system (it's the travel distance divides travel times) can reach 300–450 km/h, which realizes the comfortable travel within 3 h in medium and long distance. It complements and divides the work with WHSR and civil aviation, and realizes the rapid contact between large cities with the medium and long distance. High-density and large-scale group of point-to-point direct trains are conducted, and then magnetic intercity lines, freeways, rails, etc. are used to radiate the small and medium-size cities within the scope and collect tourists. In the city clusters, it is suitable for commuting traffic between the main node cities in the city cluster. Taking the extra-large cities of the city clusters as the core, it gathers and radiates the surrounding small and medium-size cities, forming a travel circle with business, official duties, commuting and tourism of 0.5–1 h as the main target, operating the high-density, medium-small grouped point-to-point trains at hub-station and station-station.

5.3 The Basic Structure of MHSR Train

Utilizing the electromagnetic principle of the same pole repulsion and hetero polar attraction, the medium–low speed magnetic train is lifted by the electromagnetic force and suspended on the track. After energizing, the electromagnet generates a magnetic for attracting the track, sucks up the train, and balances the electromagnetic attraction with gravity, thereby suspending the train above the track. The low-speed magnetic train has a rated suspension clearance of 8 mm, allowing passengers to experience a stable ride. When the suspension clearance is less than the rated value, the train control system reduces the electromagnetic force, and the train descends. When the suspension gap is greater than the rated value, the control system increases the electromagnetic force and sucks the train upward. This control process is carried out more than ten thousand times per second, so that the electromagnetic force keeps a dynamic balance with gravity and the train is stably suspended at a position of 8 mm from the track. The medium–low speed magnetic train is driven by electromagnetic force. When the linear motor of the train is energized, a traveling wave magnetic field is formed between the train and the track, like the wave that pushing the surface object, the magnetic train is pushed along the track.

(1) MHSR train system. The MHSR train is mainly composed of three parts: the suspension system, the propulsion system and the guiding system. Although a magnetic-independent propulsion system can be used, in most designs, the functions of these three parts are performed by magnetic force. The design of the suspension system can be divided into two directions: the normal conduction type adopted in Germany and the superconducting type adopted in Japan. Those are the Electromagnetic Suspension system (EMS) and the Electrodynamic Suspension system (EDS) from suspending technique.

① Electromagnetic Suspension system (EMS). The EMS is a suction suspension system in which the electromagnets on the locomotive and the ferromagnetic tracks on the guide rails repel each other to create a suspension. When the normal conduction MHSR train is working, firstly, the electromagnetic repulsive force of the suspension and the guiding electromagnet in the lower part of the vehicle is adjusted, and the electromagnet reacts with the winding on both sides of the ground track to suspend the train. Under the reaction of the guiding electromagnet and the track magnet in the lower part of the vehicle, the wheel and the track are kept at a certain lateral distance, and the contactless support and the non-contact guiding of the wheel rail in the horizontal direction and the vertical direction are realized.

② Electrodynamic Suspension system (EDS). The EDS system puts the magnet on the running locomotive for generating the electric current on the guide rail. As the gap between the locomotive and the guide rail is reduced, the electromagnetic repulsion will increase, so that the generated electromagnetic repulsion will provide stable support and guidance for the locomotive. The EDS system has been further developed under the cryogenics superconducting technology (Table 5.3).

(2) Superconducting MHSR train system. The main characteristics of the superconducting MHSR train are the complete electrical conductivity and complete diamagnetism of its superconducting components at relatively low temperatures. The superconducting magnet is composed of superconducting coils which are made of superconducting material, and it not only has zero current resistance, but also can transmit a strong current which ordinary wires cannot match. This characteristic enables it to be made into a small size and powerful electromagnet. Figure 5.15 shows the components of Superconducting MHSR train system.

Table 5.3 Comparison between electromagnetic suspension system and electrodynamic suspension system

Name	Electromagnetic suspension system (EMS)	Electrodynamic suspension system (EDS)
Suspension state	Since the suspension and guidance are actually independent of the train running speed, the following vehicles can still keep the suspension state even in the parking state	EDS cannot guarantee the suspension state when the locomotive's speed is below about 55 km/h
System	The suspension gap between the vehicle and the track is 10 mm, which is guaranteed by a high-precision electronic adjustment system	The locomotive must be equipped with a wheel-like device to effectively support the locomotive during its "takeoff" and "landing"

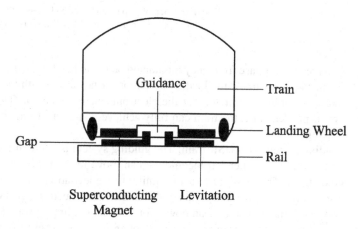

Fig. 5.15 Components of superconducting MHSR train system (picture from the network)

When three-phase alternating current is provided to the drive winding on both sides of the track, which corresponds to the vehicle speed frequency, a moving electromagnetic field will be generated, and then magnetic waves will be generated on the train guide. At this point, the superconducting magnet on the train is subjected to a thrust which is synchronous with the moving magnetic field, and this thrust pushes the train forward. A high-precision instrument installed on the ground rail to detect the position of the vehicle will adjust the three-phase alternating current supply mode according to the information from the detector, and accurately control the electromagnetic waveform to ensure the normal operation of the train.

① Propulsion system in MHSR train. The drive of the MHSR train uses the principle of synchronous linear motor. When the armature coil as the stator is energized, the rotor of the motor is driven to rotate due to electromagnetic induction. Similarly, when a substation arranged along the line

Table 5.4 Characteristics of magnetic suspension system

Index	Characteristics	Details
1.	High safety	The vehicle is suspended and orbited, and its structural form determines that there will be no derailment and rollover accidents during operation
2.	Friendly environment	Low noise, low radiation and zero emissions
3.	Good economy	Low construction cost, low maintenance and low comprehensive energy consumption
4.	Adaptable	Low floor space, small turning radius and strong climbing ability
5.	Smooth and comfortable ride	The train is in suspension and there is no contact between the body and the track, so it runs smoothly and rides comfortably
6.	Low noise	Due to the lack of impact and friction between the wheel and rail, the noise of the MHSR train is very low

provides three-phase frequency modulated power to the drive winding on the inside of the track, the loading system is pushed along with the train, like a "rotor" of the motor, for the electromagnetic induction. Thus, in the suspended state, the train can fully achieve non-contact traction and braking.

② Guiding system in MHSR train: The guiding system is a lateral force to ensure that the suspended locomotive can move in the direction of the guide track. The required thrust is similar to the levitation force and can also be divided into gravity and repulsion. The same electromagnet on the bottom of the locomotive can power for both the guiding system and the suspension system, or the train can adopt an independent electromagnet of guiding system.

(3) Characteristics of magnetic suspension system. Taking the normal guided medium–low speed magnetic suspension system as an example, the suspension and guidance of the train on the track are realized by electromagnetic force. The linear motor is used to pull the train along the track for realizing the "no-contact" motion, and the principle of "same-pole charges repel and opposite-poles charges attract" of the electromagnet is used. The electromagnetic force overcomes the gravity, so that the car body is suspended above the track, and is driven by the linear motor to drive and fly with ticking to the track. Therefore, it has the special advantage that WHSR does not have (Table 5.4).

5.4   The Main Types of MHSR Trains

The MHSR train is a new type of vehicle consisting of the non-contact electromagnetic suspension system, the guiding system and the driving system. The MHSR train can be classified as two types, i.e. superconducting and normal conducting. In short, in terms of the internal technology, there is a difference between the two types on the system whether to use magnetic repulsion or magnetic attraction. At present, there are three typical magnetic suspension technologies:

The first is the electromagnetic levitation technology invented by Germany, which is used in maglev trains under construction in Shanghai, Changsha and Beijing.

The second is the low temperature superconducting magnetic levitation technology invented in Japan, such as the central Shinkansen magnetic floating line in construction in Japan.

The third is the high temperature superconducting magnetic levitation. Unlike the low-temperature superconducting magnetic levitation which uses liquid helium cooling (-269 °C), high-temperature superconducting magnetic levitation uses liquid nitrogen cooling (-196 °C), and the operating temperature is increased. At present, the high temperature superconducting magnetic levitation technology is not mature enough, and the pilot line needs to be studied before application.

Therefore, MHSR trains can be divided into three types: electromagnetic suspension, electrodynamics suspension and high temperature superconducting suspension. It took 66 years for the invention and commercial application of the Germany's electromagnetic suspension technology. Japan's low temperature superconducting magnetic suspension took 45 years, and China's high temperature superconducting magnetic suspension took 30 years (Table 5.5).

According to different classification standards, magnetic trains can be divided into different types: Types of MHSR trains based on electromagnet type: normal conducting attraction MHSR train and the superconducting MHSR train (Table 5.6).

Types of MHSR trains based on the suspension technology: electromagnetic suspension (EMS), permanent repulsive suspension (PRS) and electrodynamics suspension (EDS) (Table 5.7).

Table 5.5 Contrastive analysis of EMS, EDS and high temperature superconducting magnetic suspension

Name	EMS and EDS	High temperature superconducting magnetic suspension
Application	Already invested or about to enter commercial applications	Experimental stage
Speed	Up to 600 km/h	Theoretical speed exceeds 600 km/h
merits and demerits	High safety, environmental friendliness, good economy, adaptability, stable ride and low noise	Simple operation, green and friendly environment but at a higher cost

Table 5.6 Comparative analysis of different MHSR trains based on electromagnet type

Name	Normal conducting attraction type	Superconducting repel type
Principle	Using a constant magnet and a guide rail as the magnetizer	Suspension operation between train and line by using superconducting magnets and low temperature technology
Suspension gap	About 10 mm	About 100 mm
Speed	300–500 km/h	The maximum operation speed can reach 1000 km/h
Applicability	Suitable for intercity and suburban transportation	It does not suspend at a low speed and only suspends when the speed reaches 100 km/h
Cost	NA	Construction technology and cost are much higher than that of a normal conducting attraction MHSR train

Table 5.7 Comparative analysis of different MHSR trains based on suspension technology

Index	MHSR train	Principle	Characteristics
1.	EMS	Using the attraction between the permanent magnetic material and the electromagnet	Most of suspension is on this way
2.	PRS	Using the repulsive force between the permanent magnets and the poles, the repulsion is 0.1 MPa	The disadvantage is the unstable factor of lateral displacement
3.	EDS	Relying on the relative motion of the excitation coil and the short-circuit coil to obtain repulsive force	Not suitable for low speed

5.5   The Development of MHSR

MHSR includes four major systems: track, vehicle, traction and operation control, with 16 core technologies. Currently, Germany, Japan, South Korea and China are the countries in the world that have a MHSR train test or operation route. Germany: the MHSR train has a 31.5 km track and runs at speeds of up to 420 km/h. Japan's JR Maglev: Japan's superconductor MHSR train is dominated by Tokai Railway (JR Tokai) and Railway General Technology Research Institute (JR Research Institute). The first experimental train JR-Maglev MLX01 was developed in the 1970s, and five experimental cars and tracks were built in Yamanashi Prefecture. On December 2, 2003, the maximum speed of it reached 581 km/h, and in 2015, it reached 603 km/h, which was setting a landed extreme speed for vehicles with carriage (Table 5.8).

For passenger transport, the main purpose of increasing speed is to shorten the travel time of passengers. Therefore, the requirements of operating speed are closely related to the length of travel distance. Various vehicles play a key role in different

Table 5.8 Operating lines of MHSR in countries around the world

Index	Country	Line	Operation mileage/(km)	Maximum speed/(km/h)
1.	Japan	Tobu Kyuryo line	8.9	100
2.	China	Shanghai magnetic demonstration line	30	430
3.	S. Korea	Incheon airport MHSR line	6.1	110
4.	China	Changsha medium–low speed MHSR line	18.55	100

travel distances according to their own speed, safety, comfort and economic characteristics. The analysis of the total travel time and travel distance of various modes of transport indicates that in terms of the total travel time, the WHSR line where the operating speed is 350 km/h is superior to the aircraft when the travel distance is less than 800 km. However, for the high-speed magnetic train whose operating speed is 500 km/h, the travel distance superior to that of the aircraft will reach above 1000 km (Table 5.9).

5.5.1 The Development Status of MHSR Train

The countries that have studied the MHSR trains in the world are mainly Germany, Japan, the United Kingdom, Canada, the United States, the Soviet Union and China. The United States and the Soviet Union abandoned their research plans in the 1970s and 1980s respectively. However, the United States recently restarted its research program (Table 5.10).

The MHSR train does not rub against the track during operating and emits less noise. Magnetic trains generally pass through the flat or over the hills with an elevated height of above 5 m, which inevitably cause the damage to the ecological environment caused by trenching. The MHSR train runs on the rails and it is charged according to the fire protection standards of the aircraft (Fig. 5.16).

The first rail appeared in 1825. After 160 years of hard work, its operating speed broke through 300 km/h. However, it took about 30 years to increase from 300 km/h to 380 km/h. With the continuous improvement and development of rail technology, the train has the potential of increasing speed. However, the cost of the 350 km/h HSR is nearly twice as high as the 160 km/h HSR and three times higher than the traditional rail of 120 km/h.

A small model of the world's first MHSR train appeared in Germany in 1969. And only a decade later, the magnetic train technology created a speed record of 170 km/h. Although the technology is not yet mature, it has entered the construction phase of 300 km/h practical operation. The results of the energy consumption study and actual test of the MHSR train showed that the energy consumption per seat kilometer of

Table 5.9 Maximum speed history of MHSR train

Year	Country	Train	Speed/(km/h)
1971	Germany	Prinzipfahrzeug	90
1971	Germany	TR-02 (TSST)	164
1972	Japan	ML100	60 (Manned)
1973	Germany	TR04	250 (Manned)
1974	Germany	EET-01	230 (Unmanned)
1975	Germany	Komet	401.3 (Propelled by a steam rocket, unmanned)
1978	Japan	HSST-01	307.8 (Propelled by a steam rocket, unmanned)
1978	Japan	HSST-02	110 (Manned)
1979	Japan	ML-500R	517 (Unmanned)
1979	Japan	ML-500R	504 (Unmanned), break 500 km/h at first time
1987	Germany	TR-06	406 (Manned)
1987	Japan	MLU001	400 (Manned)
1988	Germany	TR-06	412.6 (Manned)
1989	Germany	TR-07	436 (Manned)
1993	Germany	TR-07	450 (Manned)
1994	Japan	MLU002N	431 (Unmanned)
1997	Japan	MLX01	531 (Manned)
1997	Japan	MLX01	550 (Unmanned)
1999	Japan	MLX01	548 (Unmanned)
1999	Japan	MLX01	552 (Manned/5 vehicles as a group), recognized by Guinness world records
2003	China	MHSR SMT	501.5
2003	Japan	MLX01	581 (Manned/3 vehicles as a group), recognized by Guinness world records
2015	Japan	L0	590 (Manned/7 vehicles as a group)
2015	Japan	L0	603 (loaded mouse/7 vehicles as a group)

Table 5.10 Current status of early magnetic suspension studies of UK, Germany and Japan

Index	Name	U.K.	Germany	Japan
1.	Start time	1973	1968	1962
2.	Technical orientation	Normal conductor	Normal conducting	Superconducting
3.	Line	Birmingham airport-rail station of 600 m, 90 s, 1984–1995	Berlin in Germany, 135 km/h, 1989	Miyazaki test line, 204 km/h, 1972
				Yamanashi test line, 517 km/h, 1979
				Yamanashi test line, 550 km/h, 1997, World record

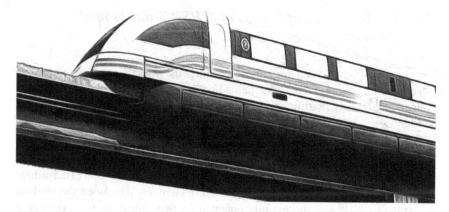

Fig. 5.16   Electromagnetic train in Germany

the magnetic train is only 1/3 of the aircraft at a speed of 500 km/h. According to the German test, when the speed of the TR MHSR train reaches 400 km/h, its energy consumption per seat kilometer is the same as that of the WHSR train with a speed of 350 km/h. When the speed of MHSR train drops to 300 km/h, its energy consumption per seat kilometer is 3.3% lower than that of the WHSR, but the construction cost is much higher (Fig. 5.17).

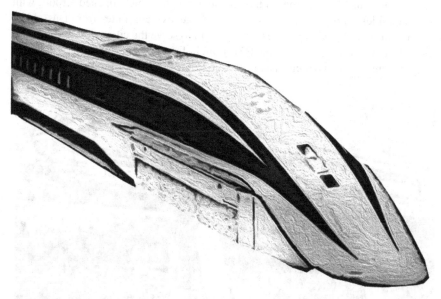

Fig. 5.17   Superconducting MHSR train in Japan

5.5.2 The Development History of MHSR Train Around the World

The promotion of high-speed magnetic levitation in the world is extremely rough, but the medium–low speed magnetic levitation lines have taken a different approach and developed vigorously in different countries, such as Germany, Japan, South Korea and China.

(1) The development history of MHSR train in Germany. Germany began research and development of magnetic traffic system around 1970 and the main development directions are as follows: the urban transportation system (Transurban) and the intercity HSR system (Transrapid). Germany developed the M-Bahn system which was successfully operated on two shorter routes: one was in Berlin and the other was at Birmingham Airport in the UK. Later they were all removed for reasons other than technology (Fig. 5.18).

Germany was trying to promote MHSR train on a series of new intercity lines. The biggest one was the construction of the new Berlin—Hamburg line. The route and station location had been carefully selected. In July 2000, the magnetic project was eventually abandoned, and the HSR technology was adopted. Nevertheless, the supporters of Maglev were still applying 6.1 billion marks ($3 billion) from the federal finances to build MHSR lines elsewhere to prove the role of the MHSR system. In the end, two lines were identified: one was in Munich, from the central city station to the newly opened airport, with a total length of 37 km; Other was the "urban express route" from Düsseldorf to Dortmund, with a total length of 78 km to serve the cities of the Ruhr area.

(2) The development history of MHSR train in Japan. Japan's research and development of the magnetic began in 1962, but their main progress on MHSR

Fig. 5.18 MHSR train in Germany

was achieved in the 1970s. Japan's magnetic levitation technology is different from that of Germany: the Japanese model uses superconducting, in which the vehicles and guides are designed based on a repulsive magnetic, while the German technology is an attractive magnetic. Repulsive magnetic technology does not work at low speed, the train body will suspend only when the train speeds up to 100 km/h with rubber tires. The double-suspension makes the vehicle more complicated, but high-speed tests prove the feasibility of this technology. In fact, Japan's MLX01 magnetic system maintains a world record for the highest testing speed of 551 km/h, and the relative speed of the two trains reaches 603 km/h.

In order to operate the magnetic levitation central Shinkansen section between Tokyo and Nagoya in 2027, Japan's JR Tokai Company conducted operation tests in the MHSR train test area in Yamanashi Prefecture on April 21, 2015. The new type "L0" train is composed of 7 coaches. On the test line with a total length of 42.8 km, the world's highest speed of 603 km/h was achieved and the train traveled at this speed for 10.8 s. The Japanese government has officially approved the construction plan of 5.5 trillion Yen (about 315 billion Yuan), while the new line is expected to finish in 2027 and extend to Osaka by 2045. The new line is officially named JR Tokai and planned to operate at a speed of 500 km/h (Fig. 5.19).

The Japanese MHSR train "Linimo" is a linear locomotive with a magnetic levitation type. It consists of three carriages with a capacity of 244 passengers and a total length of 8.9 km. There are 9 stations in total, and the maximum speed of the train is 100 km/h, which can transport 30,000 passengers per day. The operation of the MHSR train is the first implementation in Japan. It is not only the first MHSR line in Japan that is officially put into commercial operation, but also the first example of the commercial operation of the world's

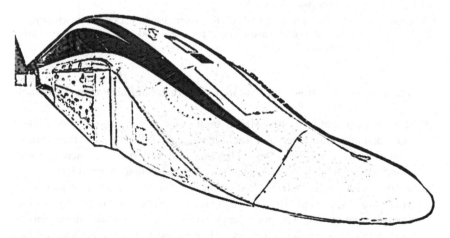

Fig. 5.19 Japan's new generation MHSR train

Fig. 5.20 Japanese MHSR train "Linimo"

Table 5.11 Features table of Nagoya—Tobu—Kyuryo Line

Line name	Nagoya—Tobu—Kyuryo Line
Line origin	The official magnetic suspension commercial line is one of the HSST plans for the Japanese express train transportation system. The line was built in conjunction with the Nagoya International Fair
Origin–destination	From the Fujioka Station of the Nagoya Subway to the Amakusa Station of the Aichi Ring Rail
Line feature	Double line, 60% of the ramps
Basic mode	HSST-IOOL type
Time to Service	On March 6, 2005, the line achieved commercial operations three years ahead of schedule
Construction cost	An average of 1 km will cost 10.3 billion yen, but it is 1/3 cheaper than the nearby Nagoya subway
Advantage	Because it is unmanned, and the vehicle and the track do not touch each other, it requires almost no maintenance and maintenance, and its operating cost can be controlled at a lower level

medium and low-speed magnetic trains by urban rail transit (Fig. 5.20; Table 5.11).

(3)   The development history of MHSR train in China. The world's first MHSR train demonstration operation line, Shanghai MHSR line, was from Pudong Longyang Road Station to Pudong International Airport. Its commercial operation began on January 4, 2003, and it was the world's first commercial magnetic suspension line. Shanghai MHSR train is a "normal conducting magnetic" (refer to as "normal guided") magnetic levitation train. It is designed by using the principle of "opposite-pole charges attract" and it is a suction suspension system that uses the electromagnet on both sides of the train, the magnets laid

Fig. 5.21 Shanghai MHSR train system

on the track, and the repulsive force generated in the magnetic field to make the vehicle suspend (using the same magnetic poles to repel each other) (Fig. 5.21).

①   Shanghai MHSR train. The Shanghai MHSR train has a speed of 430 km/h. Only one train can be allowed to operate in one power supply area. There are isolation nets at 25 m on both sides of the track, and protective equipment on the upper and lower sides. The line was built by German Transrapid Company in 2001, which was from Shanghai Pudong International Airport to Longyang Road Station. The line has a total length of 30 km and has no intermediate station. The train has a top speed of 430 km/h, an average running speed of 380 km/h and a turning radius of 8000 m. The terminal takes only 8 min.

②   Changsha MHSR train. On May 16, 2014, the first medium–low speed magnetic line with independent intellectual property rights in China: Changsha Maglev Engineering was officially started. On May 6, 2016, Changsha Maglev Express line was put into trail operation. It is also the longest medium–low speed magnetic operating line in the world. Compared with Shanghai MHSR train that was introduced from Germany and flew in the world's first commercial MHSR line, Changsha medium–low speed magnetic train has the characteristics of safety, low noise, small turning radius and strong climbing ability. Many achievements have reached the international leading level (Fig. 5.22).

     The Changsha middle-low speed magnetic project connects South Changsha Railway Station and Changsha Huanghua International Airport. The total length of the line is 18.55 km. There are 3 stations in the initial stage and 2 reserved stations. The design speed is 100 km/h.

(4)   The development of MHSR train in Korea. South Korea's MHSR train has a top speed of 110 km/h. On May 15, 2014, Korea's first self-developed commercial MHSR train was put into trial operation. The train is completely unmanned and has a top speed of 110 km/h. It is from Incheon International Airport and goes to Incheon Longyou Station. The total length is 6.1 km, and the line is expected to further expand.

     Compared with traditional light-rail trains, Korea's MHSR train has the advantages of low noise and low vibration because it does not generate orbital

Fig. 5.22 Changsha MHSR train system

Fig. 5.23 Korea's MHSR train system

friction when operating. In addition, the bogie of the MHSR train wraps the track, and it also reduces the risk of train derailment and overturning (Fig. 5.23).

5.6   Difference Between WHSR and MHSR

The maximum speed of MHSR is higher than that of WHSR, but the energy consumption and the cycle cost of the MHSR are lower than those of the WHSR, so the former attracts more passengers. Low noise and low vibration are also characteristics of MHSR. Comparative analysis of WHSR and MHSR are showed in Tables 5.12 and 5.13.

Table 5.12 Comparative analysis of WHSR and superconducting MHSR (first list)

Name	WHSR	MHSR
Speed	Recently developed very fast, compared with the magnetic suspension, the difference of running time between the two stations will become smaller	A certain advantage
Compatible	Great advantage	Poor
Cost	Low investment cost	High construction costs and unknown operating costs
Energy consumption	Low	High
Others	Such as ride comfort, running map, climbing ability, noise, etc., not the key factors	

Table 5.13 Comparative analysis of WHSR and MHSR (second list)

Contrast factor	System characteristics	MHSR	WHSR
Travel time	Max speed	420–450 km/h (261–280 m/h)	300–360 km/h (186–217 m/h)
	Accelerated speed	Higher, wider	NA
Compatible	Network interconnection	Non/single line	Good/mass network
	Utilization of existing facilities	Elevated new line, need to build stations and tunnels	New line mixed with existing lines and stations
Cost	Investment costs (dollar)	$12–55 M/km ($19–88 M/m)	$6–25 M/km ($10–40 M/m)
	Operation and maintenance costs	Uncertain	Known
	Energy consumption	Higher than HSR	NA
Other factor	Sit comfort	NA	Well
	System operation map/attraction to passengers	Excellent/especially at the beginning of the attraction of passengers because of innovation	Excellent/high network penetration rate
	Impact on the surrounding environment	Low noise and vibration	The line is on the same plane most of the time

(1)   Time costs. Comparing the maximum test speed of the MHSR with the operating speed of the WHSR, the MHSR can achieve higher operating speed than the WHSR. The difference in maximum speed between MHSR and WHSR has decreased a lot in recent years. The highest test speeds of the two trains

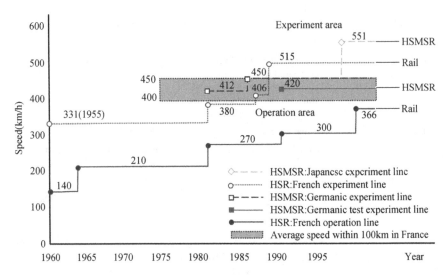

Fig. 5.24 Test speeds of MHSR trains around the world

are at the same level: the Japanese MHSR train is 603 km/h, the French TGV is 574.8 km/h, and the German "Transrapid" train is 450 km/h.

The initial acceleration of the WHSR train and the MHSR train can also be compared, because the magnitude of the acceleration is limited by the comfort level of the passenger. The MHSR train has a higher acceleration in the high-speed section than the WHSR train, so it has an advantage in the operation of longer station distances. However, in most cases, the acceleration time is a small percentage of the reduction in travel time. Therefore, although the MHSR train has advantages for the highest speed and acceleration in high-speed sections, these advantages become small at the usual station distance. Even at a distance of 100 km, the difference is only about 1 min (Fig. 5.24).

(2)   Economic cost. The construction cost of rails and stations depends on the scale of construction, mainly the grade of the track, the scale of the elevated line or tunnel. The MHSR line needs to be completely separated from the existing line, which brings a high investment in the line termination area, especially underground. Since the MHSR line has a large cross-sectional area, the WHSR can utilize the existing line in a small range in any cases. Therefore, in the same place, the WHSR is obviously superior to the MHSR one, but the maintenance cost of the MHSR is less. The biggest advantage of MHSR is that it has no physical contact with the rails, but any sophisticated equipment requires very high maintenance costs, especially the complex electronics on rails and trains.

(3)   Energy consumption. MHSR does not have wheel resistance on wheels and rails, but it requires constant energy consumption to levitate and may consume more energy than wheel-rail rolling resistance. The linear motor used in magnetic consumes more energy than the rotating motor: for example, Sky train in Vancouver and Scarborough in Toronto all use linear motor traction

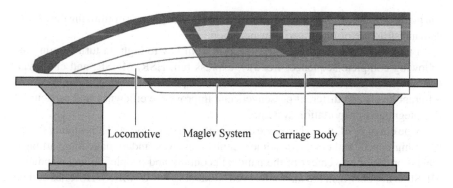

Fig. 5.25 Maglev high-speed railway system

vehicles, which consume 20–30% more energy than similar vehicles driven by ordinary rotating motors. (These two lines use wheel-rail vehicles, so there is no energy consumed by the suspension). In short, the WHSR consumes less energy than the MHSR in the same situation. The energy consumption per person kilometer of MHSR transportation is about 1/5 of that of the aircraft and 1/2 of that of the automobile. The energy consumption at the same speed is lower than that of the WHSR, and has the significant advantages of low energy consumption and low emission.

(4) Environmental impact. Since the magnetic suspension has no physical contact with the track, it has less noise and vibration than the WHSR; WHSR has its advantages in utilizing existing rail hubs in urban areas; both of them need to build tunnels in densely populated areas.

The advantage of MHSR compared to WHSR is actually very small, far less than the advantage of WHSR relative to the MHSR: especially in terms of system networking, compatibility and investment costs. However, as a new HSR system with self-contained system capability, MHSR can fill the speed gap between aviation and WHSR. In particular, MHSR can optimize the technical and economic structure of transportation mode and enrich the connotation of comprehensive transportation system (Fig. 5.25).

5.7 Summary

The speed of MHSR transportation is between WHSR and civil aviation. It not only has its own target service customer base, but also attracts and complements the target customers of civil aviation and WHSR to meet the growing speed and efficiency demand of the comprehensive transportation system. The MHSR can also increase the selectivity and substitution of high-speed passenger transportation modes, and can guide the realization of rapid traffic popularization, thereby optimizing the overall demand structure of integrated transportation.

In various passenger modes, the speed is from low to high, in turn forming the following complete sequence: freeway, express rail, HSR, MHSR, and civil aviation aircraft. This can satisfy the diversified transportation service requirements of different levels and different passengers and improve the efficiency and capacity of the integrated transportation system.

In the era of HSR, why should we develop the MHSR projects? It is mainly for the "post-high-speed rail era" to do some technical reserves, and to make more arrangements for the improvement of the national economy and people's living standards. MHSR is not only fast, but also safe, even surpassing the WHSR. WHSR has certain physical limitations, and is more suitable for speeds below 400 km/h. It is more economical and practical, but when the speed is expected to exceed 400 km/h, the MHSR is needed. To further improve the comfort of HSR and achieve the highest operating speed in the world, from the current cognition, the MHSR is the only option. The MHSR train is light enough to provide greater traction power and no vibration. Therefore, we must also develop the MHSR to seize the opportunity.

Chapter 6
Super-Speed Rail (SSR)

In 1776, there would be no steam engine without Watt's fancy. In 1879, there would be no electric generator without Edison's unthinkable idea. In 1886, there would be no train without Carl Benz's cock and boll stories. In 1903, there would be no plane without the Wright brothers' impossible task. In 2013, there would be no SSR without Elon musk's naive dream.

Now human mainly use aircraft as the high-speed long-distance passenger transport and the operating speed of the passenger aircraft is about 1000 km/h. However, for long-distance travel of more than 5000 km, the time and economic cost of traveling by aircraft are staggering, and the aircraft also caused serious environmental pollution. Especially the continuous air disasters have made people aware of the shortcomings of the civil aviation system. By this time, a new type of vehicle with a minimum speed of 1000 km/h, energy consumption less than 1/10 of that of civil aviation passenger aircraft, less noise, few exhaust pollution and lower accident rate: a super-speed rail (vacuum pipeline MHSR train) is on the horizon. Figure 6.1 shows the architecture of SSR train.

The resistance of the HSR in the high-speed travel mainly comes from two aspects: air resistance and the wheel friction. The HSR train is suspended above the ground in the magnetic floating environment, and it can avoid the wheel friction. Hence, the power of HSR train is mainly used for overcoming strong air resistance. According to the aerodynamic theory, the air resistance is proportional to the square of velocity, so the input force is proportional to the cubic of velocity. To solve air resistance problem, some scholars thought of vacuum tubes. In vacuum, SSR can run at high speed without any resistances. Figure 6.2 shows the SSR pipeline.

The several ways of modern transportation such as ships, trains, planes, cars, etc. have brought the progress and prosperity to human. However, it also has brought the pollution, traffic jam and death. SSR, which Musk calls the "fifth transportation mode", is as fast as airplanes and cheaper than train. It can operate continuously under any weather conditions and discharge no carbon emissions. The SSR turns the city to the subway site. Then the geographical boundaries will disappear and a global village under the SSR environment will come.

© Southwest Jiaotong University Press 2023
Q. Hu and S. Qu, *A Brief History of High-Speed Rail*,
https://doi.org/10.1007/978-981-19-3635-7_6

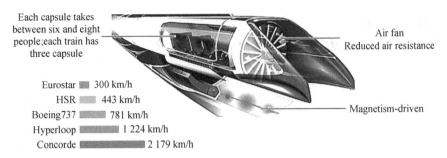

Each capsule takes between six and eight people;each train has three capsule

Air fan
Reduced air resistance

Eurostar ■ 300 km/h
HSR ■ 443 km/h
Boeing737 ■ 781 km/h
Hyperloop ■ 1 224 km/h
Concorde ■ 2 179 km/h

Magnetism-driven

Fig. 6.1 The architecture of SSR train (picture from the network)

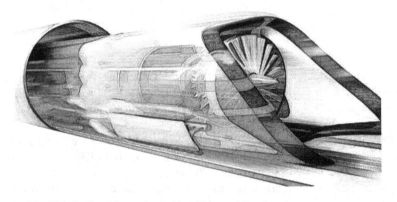

Fig. 6.2 The SSR pipeline (picture from network)

As the concept of "Super-Speed Rail" has not yet been popularized, most people are amazed at its theoretical speed, but they can't find reliable channels to understand its technical details. Meanwhile, there have been many voices of doubt in society because of a lack of scientific understanding of SSR. Therefore, based on the definition of SSR, this chapter analyzes the operation principle, basic structure and attribute characteristics of SSR, and identifies the main problems and explore the feasibility of SSR. Figure 6.3 shows the SSR station.

6.1   The Basic Principle of SSR

The SSR is a type of transportation vehicle which is designed based on the theory of ETT (Evacuated Tube Transportation). It has the feathers of super speed, high safety, low energy consumption, low noise, no vibration and no pollution, etc. It could be a new generation of transportation after cars, ships, trains and airplanes. In the future, under the premise of oil shortage, the ETT will be able to provide a popular ground super-speed vehicles to make up for the shortage of the aircraft. Therefore, the

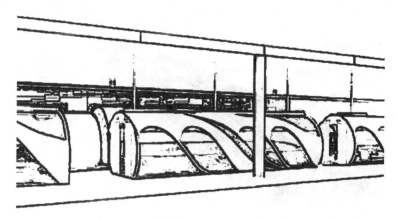

Fig. 6.3 The SSR station

significance of ETT technology is similar to that of the original steam engine which replaced horsepower and the ETT technology will bring epoch-making changes. Then civil aviation and rail transport will be replaced by large areas, then humankind will enter a cleaner, more efficient travel era. Figure 6.4 shows the imagine Figure of SSR.

(1) The principle of ETT technology. The principle of ETT technology is to build a closed pipeline on the ground or underground, and then turn it into a vacuum or partial vacuum by means of a vacuum pump. Driving in such an environment, driving resistance will be greatly reduced. In addition, the energy consumption and the aerodynamic noise can be greatly reduced, which is in line with environmental protection requirements. Vacuum tube magnetic interstellar train (VCMS) is a kind of train which has not been built yet, and it is the fastest transportation vehicle in the world. Figure 6.5 shows the frame diagram of SSR.

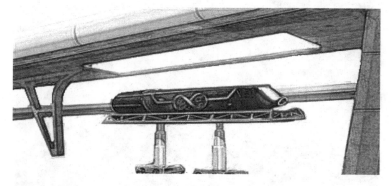

Fig. 6.4 The imagine figure of SSR

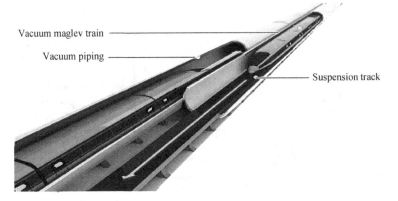

Vacuum maglev train

Vacuum piping

Suspension track

Fig. 6.5 The frame diagram of SSR (picture from network)

(2)   The features of ETT technology. The vacuum magnetic train travels in a closed vacuum pipeline. The air resistance, friction resistance and weather have no effect on it. The cost of its passenger dedicated rail is lower than that of the traditional rail. The speed of the rail can reach 1000–20,000 km/h, which is several times higher than that of plane, while the energy consumption is many times lower than that of plane. SSR may become the fastest means of transportation in the twenty-first century.

①   Low cost: The cost of ETT will be very cheap, only 1/4 of that of the freeway, 1/2 of that of the HSR.

②   Strong high environmental protection: The energy supply of the vacuum pipeline transportation can be fully supplied by solar energy without environmental pollution. Figure 6.6 shows the station map of SSR.

Fig. 6.6 The station map of SSR

③   Super speed: The speed of the SSR can reach at least 1000 km/h. It is far greater than the 400 km/h "cordon" of the WHSR. So, its speed is much higher than any other form of transportation's.

④   Good energy-saving effect: The super high-speed rail can make its energy consumption far lower than aircraft, wheel rail, magnetic high-speed rail, car, and other means of transport by eliminating the effects of air resistance.

⑤   High safety: SSR travels in the vacuum pipeline without being affected by the external environment. When the pipeline breaks and air enters, it only affects the speed of operation, which can guarantee the absolute safety of passengers.

⑥   Low noise: When the SSR train is running in a vacuum duct, there is no air as a conductor, so the operation process will not produce high decibel noise.

6.2   The Development History of SSR

The best way to ensure comfort and energy efficiency at super high speed is to run the train in a near-vacuum environment, where the idea of a vacuum tube SSR train was born (it is also called Hyperloop in U.S.). The idea of SSR was first proposed by the chief executive of Tesla, Elon Musk, in 2013. In theory, when the train is suspended above the track, it can reach ultra-speed, and the top speed is designed to be 750 miles per hour (about sonic speed 1200 km/h). The speed of the SSR is thirty-four times faster than that of the bullet train, and it can reach twice the speed of the plane. At the same time, it will be able to replenish its own energy, and when solar panels are installed in the system, the energy received will exceed the energy consumed by the entire system. In addition, the system also has an energy storage facility and the train can travel for 1 week without the battery panel. As expected, the transport system consists of low-pressure steel tubes and aluminum capsule bodies that are internally protected by gas, with a maximum operating speed of more than 745 miles per hour. Figure 6.7 shows the design of SSR from Argo Design company. The SSR cabin is shaped like a capsule, as shown in Fig. 6.8.

(1)   The train of Hyperloop. The Hyperloop train is expected to build fixed vacuum pipes on the ground that act like rail tracks, and to place "capsules" in the pipelines. According to the research team, the cabin is shaped like a space capsule. Its individual mass is 183 kg, lighter than a car. It is about 4.87 m long and can accommodate 4 to 6 passengers or 367 kg of cargo.

(2)   The operation mode of Hyperloop. The "capsule" train floats in the vacuum pipeline and the cockpit is activated by an ejection device like firing a cannon-ball, non-stop to the destination. The maximum operating speed of the "capsule" train could reach 6500 km because of the vacuum environment and no friction. In this way, the travel from New York to Los Angeles only needs

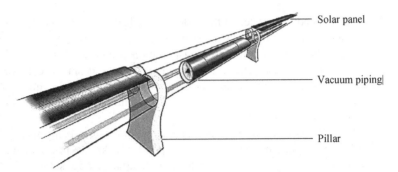

Fig. 6.7 The design of SSR from Argo Design company

Fig. 6.8 The capsule cabin

5 min, the travel from New York to Beijing only needs 2 h, the global travel only needs 6 h.

6.3   The Architecture Design of SSR

The design concept of SSR has been put forward for many years, some people put forward the idea of transportation in vacuum pipeline decades ago, but it has not been realized because of the limitation of technical conditions. In recent years, the concept of SSR was put forward again. It is a manifestation of technological progress, a further promotion of vehicle requirements and represents the pursuit of "faster". The design concept of the train and line of SSR were put forward, which represented a further effort on the commercial operation of the SSR. Figure 6.9 shows the design drawing of SSR.

(1)   Pressurization to solve operational problems. The cockpit is also called the Super-speed rail train (SSR train) or the Super train. Assuming that the departure interval of the SSR train is 2 min and the peak time can be as fast as 30 s. One SSR train needs at least 28 seats in order to meet the load of 840 people per hour during the peak hours. Now forty cockpits may be required during the peak hours, 6 of them are at the terminal for passengers to get on and off at the same time. The Super Train will have emergency brakes and engine-driven wheels, and once the massive decompression of SSR trains occurred, others trains will automatically slam on the brakes and the entire vacuum pipeline

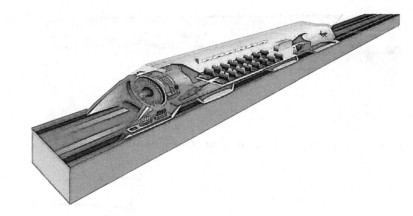

Fig. 6.9 The design drawing of SSR

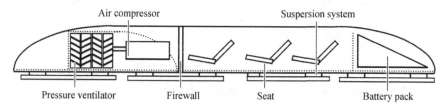

Fig. 6.10 The structure drawing of SSR train

will quickly start to supercharge. Then the super train uses the engine-driven wheels to run to safety. All the SSR trains will be equipped with a reserve of air to ensure the passengers safety under the worst conditions. Figure 6.10 shows the structure drawing of SSR train.

(2) Air cushion to solve suspension problem. At present, the method to solve the wheel friction is magnetic levitation. However, there are few commercial magnetic levitation lines in the world because of the high cost of magnetic levitation. The head of the vehicle is fitted with an engine and fan blades, a cockpit in the middle and a battery in the rear. By installing a compressed fan in the head of the train, the air is sucked in and discharged from the bottom of the train, forming a few millimeters of air cushion to make the train suspended. The forward power of the train is generated by the head compression fan. The front of the cockpit is equipped with an electric turbo compressor fan, which can send the high-pressure air in the front part into the skateboard and cabin. It can reduce the air resistance in front part of the entire vehicle, while the magnets and electromagnetic pulses in the skateboard can make the cockpit obtain the initial impetus. Figure 6.11 shows the operating diagram of the SSR.

(3) Pipe to solve air resistance problem. In order to achieve a good energy economy, the pipeline size is precise optimized to solve the air resistance problem by vacuum pumping. It is similar to an aircraft climbing to high-altitude flight to

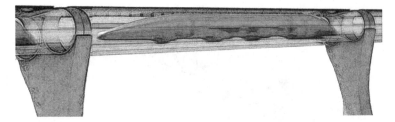

Fig. 6.11 The operating diagram of the SSR (picture from internet)

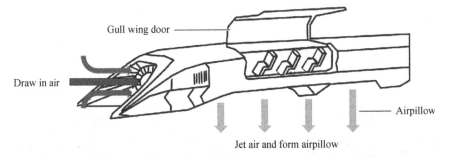

Fig. 6.12 The design principle diagram of the passenger compartment

reduce air resistance, but the cost of the vacuum is very high, and it is difficult to maintain a vacuum, because the air leakage may occur at any time. By pumping the pipe vacuum to about 10 QPA, reducing air density by reducing the air pressure in the pipeline to reduce friction consumption, the pressure is only 1/6 of that of the Martian atmosphere, which is equivalent to that of the air pressure when the plane flies at the height of 15,000 ft.

The passenger compartment is shown in Fig. 6.12, each streamlined seal cabin can accommodate 28 passengers, and one capsule can be issued in a few minutes. Two passengers are in a row, the luggage are concentrated in the front of the seal cabin or tail. As the system accelerates, passengers are subjected to less than 0.5 g (half the force of gravity).

6.3.1 The Design Concept of the SSR Train

The shape and interior of the SSR train will be similar to that of the traditional rail train, but the volume will be smaller than that of the traditional HSR trains, and even smaller than that of modern subway trains. For the vacuum pipelines operated by trains, experts are more inclined to the inner layer with steel pipe and the outer layer with reinforced concrete construction, which is mainly to reduce the amount of steel pipe and save costs. The vacuum pipeline is designed to have two doors at all the

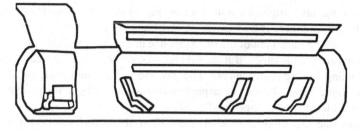

Fig. 6.13 The planning drawing of cabin of the SSR train

Each capsule carries 6-8 people,and each train consists of 3 capsule

The front blades draw air into the back of the train,reducing air resistance

Pulled by magnetic force

Fig. 6.14 The 3-dimensional form of the SSR cabin

entrance and exit. When the train is running, the staff firstly opens the outer door. The train will enter the inter layer of the pipeline which is between two doors from the station. When the outer door is closed, the vacuum pump starts to take away the air. At this time, the staff opens the inner door, the train will enter the vacuum pipeline and start to accelerate and run. And the order is reverse when the train is out of the pipeline, firstly the inner is door opened, secondly the train runs out, then the inner door is closed, finally the outer door is opened. The above process is similar to the operation mode of astronauts entering and leaving space in space (Fig. 6.13).

The vacuum MHSR train will be more stable than the plane when running. Although it is operated in a vacuum environment, the cabin is definitely not a vacuum environment. The fully sealed cabin will simulate the daily train environment and make passengers feel comfortable (Fig. 6.14).

6.3.2 The Design Concept of the Super Line

The super line is made up of pipelines. The super line requires a pumping station to be set up at the pipeline. A pumping station should be set up every 2 or 3 km in the vacuum pipeline, to extract air from the pipe. According to the design standards, the pressure in the pipeline should even reach 0.001 air pressure, that is, 1 thousandth of the atmospheric pressure, such pressure range is also the basic guarantee of the operation of trains at a high speed. The joint between the pipelines, must be sealed securely. In addition, there are many pumping stations along the pipeline, and the openings should be reserved for maintenance, inspection and emergency conditions.

These openings are airtight when the vacuum piping system was normally operated and it must ensure there is no leakage. A few leaks are unavoidable in the air lock part of the vehicle in and out of the main pipeline at each station along the line when the system runs continuously. But the seal must be reliable when closed to meet the corresponding sealing requirements. The pipeline is a vacuum state, in which the magnetic vehicle must be the atmosphere environment suitable for people, so the vehicle must have good sealing conditions.

The pipeline is composed of steel. Each 30 m of pipeline was propped by one holder. The structure is very strong with a certain seismic effect. Meanwhile, the surface of the pipeline is covered with solar panels to power the system. The power of the entire SSR train system is 21 MW and the surface-covered solar panel can provide 57 MW, which is fully enough. In order to avoid the discomfort caused by high-speed turning, the route choice is to try to maintain a straight line (Figs. 6.15 and 6.16).

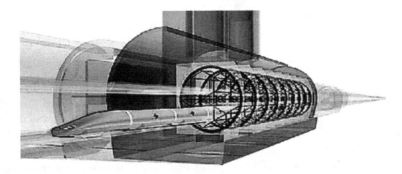

Fig. 6.15 The imagine figure of the SSR line

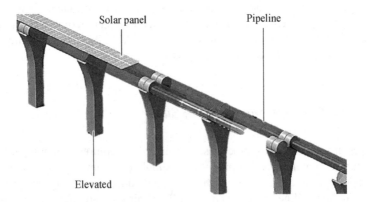

Fig. 6.16 The schematic diagram of the SSR pipeline

6.4 The Definition of SSR

Why is the SSR known as the "the fifth mode of vehicle" and what are the salient features of it that compared to other types of transportation? How is it different from other modes of transportation in operation? What is its specific definition? What is its basic characteristic?

6.4.1 The Basic Definition of SSR

The SSR is a kind of transportation, which is designed by the theory of "vacuum pipeline transportation". It connects a series of "vacuum pipelines" to form the whole transport line system, so that the passengers can arrive from A to B in a few minutes. As a means of transportation, the SSR save travel time and improve the delivery efficiency, it is convenient for passengers. Figure 6.17 shows the imagine Figure of the SSR train.

The SSR train is a kind of super train which runs in the vacuum pipeline, and it belongs to the Evacuated Tube Transport and travels in the airtight vacuum pipeline. SSR is unaffected by air resistance, friction and weather, especially unaffected by natural environment (such as gale, rainstorm, debris flow, low temperature, etc.). The super train can be regarded as an ideal mode of transportation because it can reach 1000–20,000 km/h, which is several times faster than aircraft. The advantages of SSR are no need for on-board power, high security, low energy consumption, static suspension, low noise, not easy to tinnitus, light body, and high-frequency departure; The disadvantage of SSR is the greatly increase in the cost of rail.

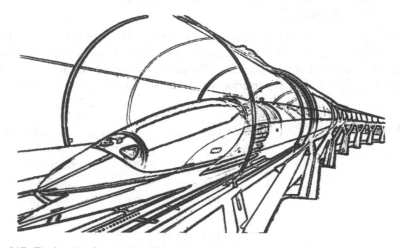

Fig. 6.17 The imagine figure of the SSR train

The Relationship Between Velocity and Air Resistance

Operating in a dense surface atmosphere, high-speed vehicles inevitably suffer friction (including contact friction and air friction, mainly air friction) resistance. The maximum speed of the surface vehicle is about 500 km/h, while the theoretical maximum velocity in the piping transport system can reach above 20,000 km/h. The resistance mainly comes from the air when the HSR exceeds the speed of 300 km/h, and the resistances can be up to 90% when the HSR exceeds the speed of 400 km/h, 99% when it is up to 500 km/h. The speed of HSR cannot exceed the speed of airplane, mainly because the surface air resistance encountered by the train is much greater than the air resistance encountered by the plane. Therefore, the speed of vehicles is related to air resistance, the greater the air resistance is, the smaller the speed of transport is.

(1)   The resistance relationship of the surface. When the vehicle is running on the ground, it faces 1 bar pressure. The speed of the rail train is the king of the surface compared with the cars, ships and ordinary trains. According to the latest research, the highest running speed on the ground HSR (wheel-rail) in normal operation is 400 km/h, and the highest running speed of the ground HSR (magnetic-rail) in normal operation is 500 km/h. Therefore, on the surface, no matter which means of transport, the fastest running speed cannot exceed 500 km/h due to air resistance, the best operating speed at different distances is shown in Table 6.1.

(2)   The resistance relationship in air. Because the air density at different altitude is different, the air resistance at different altitude is different, and this air resistance is also associated with atmospheric pressure. Therefore, the speed of transport at different atmospheric pressure is different. According to the current research results, the optimal velocity at different atmospheric pressure is shown in Table 6.2.

Through the analysis of air resistance on the ground surface and in the air, we drew the following conclusion: for the thinner the air and the smaller the air resistance, the greater the speed of transportation. Therefore, if a pipeline is constructed and the air in it is discharged, the interior of the pipeline becomes vacuum, so that the

Table 6.1 The optimal velocity value at different distances

Number	Distance/km	The running speed/(km/h)	The traffic tools
1.	<200	200	Motor train
2.	200–400	400	WHSR train
3.	400–600	500	MHSR train
4.	600–1500	1200	SSR train
5.	1500–10,000	2000	
6.	15,000–20,000	6500	
7.	>10,000	20,000	

Table 6.2 The optimal velocity value at different atmospheric pressure

Number	Height/m	Atmosphere/Pa	Flight speed/(km/h)	The running speed of The SSR/(km/h)
1.	<1000	1	400–500	<500
2.	1000–4000	0.8–1	500–600	500–1200
3.	4000–10,000	0.5–0.8	600–8000	
4.	10,000–20,000	0.2–0.5	800–1000	1200–2000
5.	12,000–15,000	0.05–0.2	1000–2000	
6.	15,000–20,000	0–0.05	2000–10,000	2000–20,000
7.	>20,000	0	>10,000	>20,000

vehicle running in the pipeline will not be subject to air resistance, and its speed can reach more than 6500 km/h; Even if there is a little air in the pipeline, as long as it is at a pressure of less than 0.1 atmospheres, the vehicle can also reach an speed of 1000–2000 km/h.

The Main Problems in the Design of SSR

In the dense atmosphere of the ground surface, the transportation of vehicles is affected by contact friction and air friction, while the main limitation of transportation is air friction, i.e. air resistance. How to improve speed of a train? The only way is to reduce friction and reduce drag. On the one hand, for contact friction, the SSR essentially use magnets to provide the thrust, relying on compressed air to provide lift, so the SSR will not have the friction resistance between wheels and rails. On the other hand, for air friction, the SSR aims to achieve the target speed, so the driving track must maintain low pressure to reduce the resistance between the SSR train and air.

(1) Design principle-air resistance problem. The SSR pipeline can be built into a closed pipe. The air inside the pipe can be ruled out and the pipe becomes a vacuum or a partial vacuum during operation. In this way, the SSR runs in a closed pipe without air resistance, the resistance of the SSR will be greatly reduced, while the energy consumption, the aerodynamic noise and the vibration of SSR train are greatly reduced as well. Figure 6.18 shows the pipeline starting point of SSR.

(2) Design principle-contact friction problem. The friction resistance of the SSR comes from air friction and contact friction. In addition to eliminating the resistance from air friction, another highlight of the SSR is suspension technology. The problem which suspension technology wants to solve is the contact friction resistance. The suspension technology uses magnetic levitation technology to

Vacuum piping   Launch   Luggage compartment Cabin

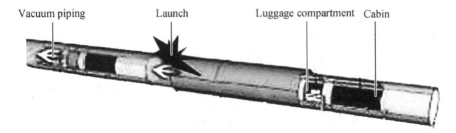

Fig. 6.18 The pipeline starting point of SSR

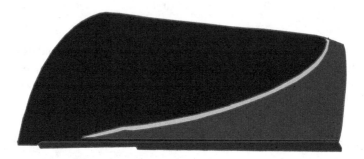

Fig. 6.19 The pipeline diagram of SSR

make delivery vehicles operate in the vacuum pipeline without contact and friction, to achieve point-to-point transport, so there is no contact friction problem. Figure 6.19 shows the pipeline diagram of SSR.

(3) Design principle-power drive problem. The SSR can be designed with self-power supply. According to U.S. expert Elon Musk, laying solar panels on the top of the pipeline can generate enough power to maintain its normal operation. When the SSR system is fitted with solar panels, the energy obtained can bear the power consumption of the entire system. Additional energy storage facilities in the SSR system can store excess energy for emergency use in the SSR system (Fig. 6.20).

Therefore, based on the above three principles, the construction of SSR is theoretically simple. Firstly, the air is pumped out of the closed environment to form a vacuum environment; Secondly, the friction is eliminated so that the delivery vehicle can be suspended in the pipeline and move at a high speed with less energy. Finally, under the impetus of solar energy, the SSR can operate fast in the vacuum pipe line.

The Basic Definition of the SHSR

The SSR system is a pipeline that is insulated from the outside air, which is pumped into a vacuum, where a MHSR train and other vehicles run (Based on Elon Musk

Fig. 6.20 The schematic diagram of solar panel of SSR (picture from network)

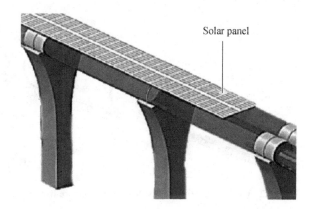

Solar panel

of the United States, the schematic diagram of the SSR is shown in Fig. 6.21). The means of delivery (i.e. the Super train) is in an almost frictionless environment, using a floating cabin to deliver passengers at 1200 km/h in a low-pressure tube. The characteristics and types of SSR are shown in Table 6.3.

(1) SSR train. The SSR is a new means of transportation built by the concept of "vacuum piping". The vehicle is a new generation of transportation after automobiles, ships, trains and airplanes, with the characteristics of ultra-speed, high safety, low energy consumption, no noise and zero pollution. Because of the vacuum operation in the pipeline and the use of magnetic levitation

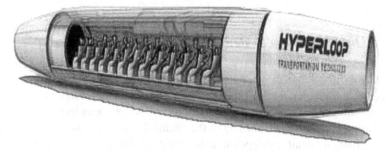

Fig. 6.21 Imagine figure of SSR train (Hyperloop) by Elon Musk (picture from network)

Table 6.3 Characteristics and types of SSR

Number	The specific characteristics	Type
1.	Transport in the pipeline	Pipeline
2.	Magnetic technology used	Rail
3.	The transport capacity is equivalent to the capacity of bus	Road
4.	Running at the same speed as the flight speed	Air
5.	Floating in the air	Water

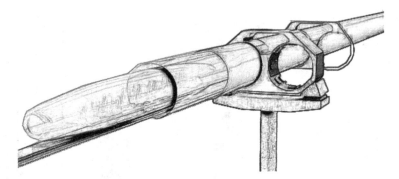

Fig. 6.22 The imagine figure of SSR pipeline (picture from network)

technology, this book suggests that the vehicle should be called vacuum rook or SSR train.

(2) Vacuum piping. The SSR is different from the traditional rail, and it is a vacuum suspension friction-free flight system, which is a new high-speed transportation system. The SSR system consists of transportation pipelines, manned compartments, vacuum equipment, suspension components, ejection and braking systems and so on. On the one hand, the SSR train is suspended inside the pipeline, and the speed can reach more than 1000 km/h; On the other hand, magnetic technology is used to float super trains in vacuum-treated pipelines, and then ejection devices are used to launch super trains along the pipelines to destinations uninterruptedly (Fig. 6.22).

6.4.2   The Propulsion System of SSR

The SSR runs in the tunnel, is not affected by the weather, and its propulsion system is placed in the pipeline without human error. Therefore, the design of the SSR is safer than the general mode of transportation. If a serious accident occurs during the operation of the SSR, passengers in the enclosed cabin are likely to be oxygen-deficient. To guard against this potential hazard, the SSR provides oxygen masks.

(1) The propulsion system of SSR: it is mainly by air compression. The SSR train includes three processes throughout the course of the journey: acceleration running at a high steady speed slowing down to the stop. And the basic requirements of the SSR propulsion system are as follows: during the start-up phase, the train can accelerate the cockpit from 0 to 480 km/h at a relatively low speed, which requires a large starting acceleration; In the linear acceleration region, the train can accelerate from 400 to 1220 km/h with an acceleration of 1 g (9.8 m/s^2); The fan of SSR train's head is sufficient to provide the force to maintain 1200 km/h. As for the acceleration and deceleration process, the force is completed by a linear motor on the wall of the pipeline, which accelerates

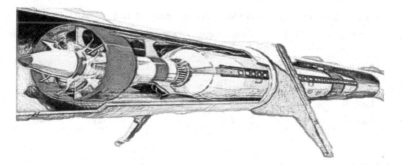

Fig. 6.23 The propulsion system of SSR

the train when it starts off and slows down when it arrives. Figure 6.23 shows the propulsion system of SSR.

(2) Energy supply pattern of SSR: solar energy is the most suitable power source for SSR. The batteries in the car are mainly used to make their own air cushions. In tunnels, an external linear motor is installed to replenish the train's power every 110 km. The pipes are covered with solar panels, which converts solar energy into electricity to make the system self-sufficient and even surplus. The power of the entire SSR train system is 21 MW, and the surface-covered solar panel can provide 57 MW of electricity, which is sufficient.

6.4.3 The Technical Characteristics of SSR

Vacuum MHSR trains are considered to be the fastest means of transportation in the world. In fact, the SSR train is a vacuum MHSR train. In addition to the superiority of the speed, the vacuum piping system also has the characteristics of fast, punctuality, large volume of traffic, comfort, safety, all-weather operation, energy saving, environmental protection and so on.

The Safety of SSR

(1) No matter which kind of transportation, safety is the first. Therefore, safety and reliability are the primary factor of passengers' travel considerations. The SSR connects many cities with huge, near-vacuum pipelines, forming a SSR network to facilitate quick travel. What about the safety of the SSR?

(2) Objective safe level. Objectively speaking, the natural environment has a large impact on various modes of transportation, but it has no effect on the SSR because it operates in a fully enclosed system, as shown in Table 6.4.

Table 6.4 Comparison of safety about SSR with other transportation modes

Type	Earthquake	Wind	Temperature	Thunder and lighting	Rainstorm	Debris flow
SSR	Smallest	Smallest	Smallest	Smallest	Smallest	Smallest
Plane	Small	Small	Smallest	Largest	Small	Small
Train	Largest	Large	Large	Large	Largest	Largest
car	Large	Largest	Largest	Small	Large	Large

Table 6.5 Subjective safety of SSR

1.	With speed limits, the cabin is fully contained in the pipeline and is not easily derailed
2.	Have advanced control and guarantee system, it is not easy to appear man-made accident
3.	There is a safety compartment along the vacuum pipeline, which can be escaped from the safe cabin in case of malfunction

(3)   Subjective safe level. Subjectively, most of the traffic accidents are related to people, while the SSR is mainly intelligent control, less human-made effect. The subjective performance of super high speed railway is shown in Table 6.5.

In short, SSR is the safest mode of transportation, both subjectively and objectively, compared to the vehicles such as cars, airplanes, ships and WHSR.

The Comfort of SSR

Comfort is another key reason for people to choose the SSR to travel. It has been suggested that SSR is faster than airplane and cannot be tolerated by humans, but in fact they can be tolerated according to the scientific analysis of the human body. Scientific analysis shows that the human body can withstand. The reasons are as follows:

(1)   The adaptability of the human body compared with the automobile. The limit that the human body usually withstands is the acceleration around 50 m/s^2 while the hundred kilometers accelerate of a car is about 10 s. If you can easily speed up to 1000 km/h in 1 to 2 min, it is not a problem for the human body to withstand the speed of SSR.

(2)   The adaptability of the human body compared with the aircraft. The acceleration of the SSR reaches the acceleration of the aircraft, and the acceleration of the aircraft is generally 0.5–0.6 g, equivalent to the acceleration at 5 m/s^2. In fact, the acceleration of the people's 100-m sprint is much greater than this. Therefore, this acceleration is entirely within the acceptable range of human beings. The train will not always be in accelerating motion after accelerating to a certain value, and it will enter a uniform motion state. At that time, passengers do not have any sense of speed, just as smooth as the astronauts flying in the

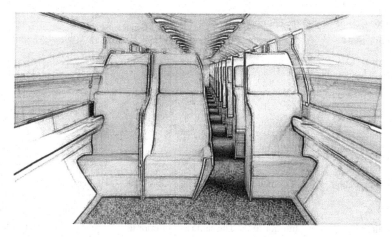

Fig. 6.24 The seat diagram of SSR

air. So even the acceleration of the SSR train is too fast, people still can bear it. Figure 6.24 shows the seat diagram of SSR.

Three performances of SSR comfort are as follows: ① Each passenger compartment of the SSR train is pressurized, the SSR train is fitted with oxygen masks and an emergency braking system to avoid the possibility of physical discomfort; ② The SSR train is ejected at the starting point. It will rely on the magnetic force to run all the way and will not encounter turbulence like an airplane on the way; ③ When the SSR train is launched, the passengers will feel the acceleration, and once the super train is full ahead, the passenger won't feel anymore. As a result, the passenger experience is comfortable on the SSR, and the SSR will be more comfortable and quieter than the WHSR and the aircraft. Especially from the technical level, the SSR comfort also includes vibration, temperature, noise, air, light and other factors. A comparison of the comfort of various vehicles is shown in Table 6.6.

The Economics of SSR

The economics is one of the main factors to be considered in building SSR. According to Musk's design philosophy, the SSR is a very economical way of transportation for any two big cities with distances not exceeding 1500 km. For example, the cost of building a SSR between Beijing and Shanghai is 6 billion Yuan. The SSR is dispatched every 3 min. Each SSR train carries 30 people, and the operating cost per trip is about 200 Yuan. Therefore, the one-way fare can be set at 200 Yuan per sheet, which is very cheap and acceptable to passengers. The comparison of the cost and operating cost of the SSR and other transportation modes is shown in Table 6.7.

(1) The economic analysis based on the cost of construction. The operation route of the SSR is the pipeline. The pipeline is supported away from the ground by

Table 6.6 Comparison of the comfort of various vehicles

The traffic tools		SSR	Traditional rail	WHSR	Cars	Aircraft
In-car stability	Longitudinal stability (°)	1.5	3	2	3	3.2
	Horizontal stability (°)	0.2	2.2	2	2.5	2.6
	Vertical stability (°)	1	2.5	2	2.8	5
In-car noise (db)		50	70	65	76	80
In-car temperature		Automatic temperature control	Higher than normal temperature	Automatic temperature control	Higher than normal temperature	Self-adjust the normal temperature
In-car air		Inferior to outdoor	The same as outdoor	Inferior to outdoor	The same as outdoor	Inferior to outdoor
In-car light		Automatic light control	The same as outdoor	Inferior to outdoor	The same as outdoor	Automatic light control

The smaller the stability value, the more stable the car environment, the more comfortable, the international stability of the value of the threshold is 2

Table 6.7 Comparison of the cost and operating cost of SSR and other transportation modes (regard the cost of SSR as 1)

Mode of transport	Cost	Operating cost	Comments
SSR	1	1	The cost of SSR is 1
Freeway	4	NA	
WHSR	10	2	

the elevated pillar. It reduces the occupation of land resources. The vacuum pipelines between two cities are built on the ground like HSR. Wherever there is a road, there can be two pipelines for two directions. And the vacuum pipeline may be "attached" to the already built high-speed bridge, thus saving the route resources and infrastructure cost. Therefore, the construction cost of SSR is lower than that of other transportation.

(2) The economic analysis based on operating cost. The use of solar power in the SSR train greatly reduced the cost of transportation. The SSR train can use its own technology to carry out multiple energy storage. After the SSR system accelerates the SSR train to a certain speed, the SSR train can rely on inertia to operate in a vacuum pipeline without any additional energy. The existing kinetic energy of the SSR train can be recovered and reused by the motor when the passengers are about to arrive and the train needs to slow down, so that the transportation cost of the SSR train is only 1/10 of the transportation cost of the WHSR train. As a result, the operating cost of the SSR is lower than that of the WHSR (Table 6.8).

Table 6.8 The cost of SSR

Type	Cost (Yuan)/100 million	The specific cost (Yuan)/100 million			Cost per kilometer (Yuan)/100 million
		Elevated cost	Land	Pipeline	
Passenger SSR	360	153	60	39	About 0.1
Passenger and cargo dual-use SSR	450				

Table 6.9 The convenience of SSR

Number	Convenience	Reason
1.	No waiting time	No need to reserve the seat and the short departure interval
2.	Everyone is equal	No grade difference of the seat
3.	Saving time	The station is located in the center of the city without transfer
4.	Choose the travel time by yourself (optional trip time)	Run automatically without fear of delay
5.	Free choice	Select running speed according to line length

The Convenience of SSR

Convenience is one of the choice conditions for SSR travel. If the SSR network can be built around the world, it will take hours to travel around the world to achieve the goal of global day trips (morning go and evening back, global work). According to Musk's design concept, the convenience of SSR is shown in Table 6.9.

The Energy Saving and Environmental Protection of SSR

In addition to the superiority of speed, the vacuum piping system of the SSR is more energy efficient and environmentally friendly, especially in low-carbon emission, energy saving and environmental protection. On the one hand, as a transportation vehicle, SSR not only has zero carbon emissions, but also no dust, fumes and other exhaust pollutants. On the other hand, vacuum pipeline transportation is a transportation method without air and friction, and it is quieter than the WHSR and the aircraft.

(1)   Analysis of the technical characteristics based on energy consumption. Due to the reduction of contact friction and air friction, the vacuum pipeline transportation consumes less energy than any traditional transportation. The transportation capacity per kilowatt-hour of the SSR is 50 times of that of the WHSR. The SSR system will be powered by solar energy. It can replenish its own energy,

Table 6.10 Comparison for energy consumption of various vehicles (kg/person)

Vehicle	Traditional rail	WHSR	SSR	Car	Aircraft
Per person equal mileage energy consumption	1	0.5	0.1	6	4

Table 6.11 Comparison of various vehicle environmental protection

Vehicle	Traditional rail	WHSR	SSR	Car	Aircraft
CO_2 emissions per person per kilometer/[mg/(km·person)]	1	0.5	0.2	10	4
Noise per person per kilometer/[db/(km·person)]	0.1	0.05	0.01	1	1

and the system has a facility to store energy. It can travel for a week without using a panel. According to the analysis of existing research results, the energy consumption comparison of various vehicles is shown in Table 6.10.

(2) Analysis of the technical characteristics based on Environmental Protection. The SSR is more than twice as fast as the aircraft, but it consumes less than 1/10 of the energy of the civilian airliner. The noise, exhaust pollution and accident rate of the SSR are close to zero. In particular, the pipelines of SSR are built underground or on the ground, there is basically no pollution to the environment. According to the analysis of existing research results, the environmental protection comparison of various modes of transportation is shown in Table 6.11.

6.4.4   The Existing Problems of SSR

Once the SSR system succeeds, it will completely subvert the human's perception of traffic. But the higher the speed is, the greater the risk is, and if an accident happens, the vacuum pipeline will bring an unthinkable disaster to passengers. At present, from a technical point of view, a variety of key technologies used in the SSR system (including low-voltage pipelines, compressors, solar energy and other technologies) are mature and feasible, but from the application level, there are many other problems to be solved (Fig. 6.25).

(1) From the view of theoretical point, the SSR system is completely feasible. In the theory, pipeline transportation is currently the most efficient and energy efficient transportation method. Vacuum MHSR train is the fastest vehicle in the world, which has been verified theoretically. Therefore, from the theoretical point of view, the construction of SSR system is entirely feasible. The specific reasons are shown in Table 6.12.

(2) From the view of application point, the construction of SSR system is very difficult, and it will not be completed in a short time. At present, the realization of

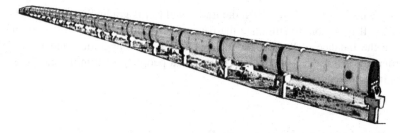

Fig. 6.25 The real figure of SSR pipeline in experiment

Table 6.12 The reason that the SSR work

Number	Feature	Reason
1.	Can realize high-speed operation	Not affected by friction and air resistance, the maximum speed can reach 20,000 km/h theoretically
2.	High safety of the system	SSR trains in fully enclosed environments are completely unaffected by weather changes
3.	Reasonable energy supply	Self-powered design, laying the solar panel above the pipeline can generate enough electricity

vacuum transportation over 1000 km is restricted, especially in technology, cost and management. Therefore, the construction of SSR system is not feasible. The reasons are shown in Table 6.13.

In short, SSR has many advantages, perhaps in the future can lead to a revolution in the field of transport and promote the great progress of human society. However, many problems in terms of technology and cost require scholars to constantly study and discuss. The development of the super HSR must be gradual, otherwise human

Table 6.13 The reason that the SSR does not work

Number	Reason	The specific performance
1.	Long distance vacuum pipe is difficult to construct	The pressure difference inside and outside the vacuum pipe is great, and the existing technical means are difficult to meet the requirement
2.	Magnetic technology is not perfect	At present, magnetic technology is not mature in practical application
3.	The problem of voltage stability	The problem of "vacuum breakdown" is easily produced in the vacuum environment, and has not been solved
4.	The problem of system managing	Transnational transport, coordinated management of different countries is difficult to solve

will pay a heavy price. Only when the traditional rail is gradually transformed into WHSR will it be able to stimulate the pace of SSR. At that time, it is possible to achieve the balance of "speed and safety" and global integration under the SSR environment and realize the global day tour that people go out in the morning and return in the evening.

6.5   Development Vision of SSR

It is a long way to realize the ultra-speed research. When the train runs, the pressure inside the pipe is 10 times lower than that outside, and the train can use more power to drive the vehicle speed forward. In 1992, researchers deduced from high-speed spin experiments that the experimental speed of HTS (high temperature superconducting) Maglev was up to 3600 km/h, but this is just theoretically, there is a long way to go.

On August 30, 2017, China Aerospace Science and Industry Corporation (CASIC) announced the launch of the "High-speed flight (HSF) train" research and demonstration. It is proposed to develop a new generation of vehicles through commercialization and marketization, combining the supersonic flight technology with the rail transit technology. Utilizing the superconducting magnetic levitation technology and the vacuum piping, researchers are committed to achieving the supersonic "near-earth flight", whose theoretical maximum speed can be achieved 4000 km/h. The HSF train is a transportation system which reduces the air resistance by using low vacuum environment and supersonic shape. It reduces the frictional resistance through magnetic levitation, and realize supersonic running. Compared with the traditional WHSR train, the speed of the HSF train is increased 10 times; Compared with the existing civil aircraft, the speed has been increased 5 times and the maximum speed can be achieved 4000 km/h. It is a great step forward for the ultimate pursuit of vehicle speed. The advent of a new era is often accompanied by a great change in transportation mode, and the CASIC HSF train project will open a new era! Fig. 6.26 shows the concept Figure of the HSR train.

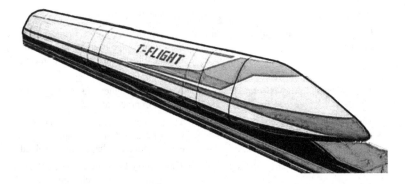

Fig. 6.26   The concept figure of the HSR train

(1)   Bright future. Vacuum tube MHSR is the most natural and direct scientific inference result based on all known traffic modes. The friction resistance between the train and the track is eliminated, and the friction resistance between the train and the atmosphere is eliminated. The ideal combination of MHSR and vacuum pipeline concept is the choice of future traffic, and the real call of the global village era. Now the era of the internet has enabled information to travel on earth at the speed of light, so that anyone in every corner can instantly know the knowledge he needs. However, the physical transmission is still limited to the time and space background. The concept of magnetic levitation in vacuum pipeline emerges as the times require, its superiority is also obvious, such as high speed, convenience, high efficiency, energy saving, clean, environmental protection, and safety. It will realize the space travel on the earth and bring the human travel to a new stage.

(2)   Complicated ways. There are still many insurmountable obstacles in the actual operation of MHSR train, let alone the vacuum transportation based on suspension technology. There are many specific problems in the practice of vacuum pipeline transportation, and how to solve these problems has never been proposed. For example, the voltage in the vacuum environment is prone to "vacuum breakdown" phenomenon, resulting in a self-sustaining discharge and the damage to the electrodes caused by the transport system paralysis. How to ensure voltage stability in a vacuum environment? In addition, the pipeline is a vacuum state, and the maglev vehicle running in it must be suitable for the human ride in the atmosphere. So how to ensure that the internal and external environment to meet the standards is also a difficult point. At present, no one can determine how much the feasibility of the vacuum pipeline transport, because all the description of the scheme is not detailed enough and short of practical necessary arguments, it is difficult to distinguish its technical rationality and engineering feasibility. As a new concept transportation system, which was put forward in 2013, SSR has been paid much attention since emergence. This concept has evolved over several years leading to several mature ideas. These ideas are the foundation of the SSR study in the future.

Vision is always ahead of reality and determines reality. The development of fast, convenient, efficient energy-saving, clean, environmental, safe, economic and practical transportation system is the goal, let us bravely to promote the future! Super-speed rail set sail!

6.6 Summary

"The concept is feasible, the theory is defective". Under the promotion of WHSR, the rapid formation of national metropolitan areas greatly narrows the distance between urban and rural and accelerates the integration of urban and rural areas. But in the future, with the promotion of the SSR, the world economic circle will also be rapidly formed. The HSR will greatly shorten the distance between countries and promote the rapid development of countries, forming a global village under the HSR environment. However, the rail is a high cost, high investment infrastructure projects, and the SSR is a high-risk facility, directly related to a country's livelihood. In the theory, it is possible to achieve higher speeds in the vacuum environment, but the maximum speed is related to not only the vacuum, but also the technology of suspension guidance system, traction system, rail system and operation control system. Therefore, the super high-speed rail also needs further theoretical research.

"The ideal is feasible, the technology is not pefect." The speed enhancement, which shortens the time and space and narrows the distance, has reconstructed the world space–time layout and realized the global integration. However, the SSR provides people the imagination space to continue to improve the travel speed, and it is hard to achieve engineering applications; there are many problems to be solved, human need to continue to study, so the myth will become reality.

Chapter 7
Global Village in High-Speed Rail Environment

In the seven continents of the world, except for Antarctica, there are the HSR applications. The development of HSR also promotes regional integration. The high compatibility of HSR is the basis of its international interconnection. Various types of HSR promote the integration of different regions. The WHSR promotes regional integration, and the MHSR promotes the promotion of the continents village, the SSR promotes global integration which is called the global village. Therefore, we must pay attention to the development of HSR. Whichever country has the core technology of HSR (WHSR, MHSR and SSR), the country owns the world.

Since the birth of HSR, the HSR experienced "four times of development", "three times of leap". At present, many countries and regions in the world have planned, built and operated high-speed rail.

Compared with other modes of transportation, HSR has strong conveying capacity, fast speed, good safety, highly punctuality rate, low energy consumption, less impact on the environment, land conservation, comfort, convenience, considerable economic benefits and good social benefits. Moreover, with the unique technical advantages to adapt to the new needs of modern social and economic development, HSR has become an inevitable choice for the development of countries around the world. The development and operation practice of China's HSR shows that the HSR has a great development space and potential in China, so we should make full use of the latecomer advantage to realize the leaping development of China's HSR. The HSR has created a new growth point for urban development, promoted the integration of central cities and satellite towns, enhanced the radiation effect of central cities on surrounding cities, and strengthened the "the same city effect" of neighboring large cities. In particular, with the development of HSR in various countries around the world, in the near future, the HSR will become the main ground transportation mode connecting countries in various regions, and will open the "global village" era together with air routes. Traveling around the world on HSR will become tourism fashion. Therefore, with the continuous improvement of HSR technology, the HSR between countries and regions around the world is like a bus, with many shifts and short intervals, without the congestion of freeway and the delay of air plane,

© Southwest Jiaotong University Press 2023
Q. Hu and S. Qu, *A Brief History of High-Speed Rail*,
https://doi.org/10.1007/978-981-19-3635-7_7

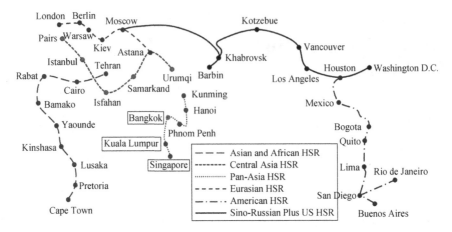

Fig. 7.1 The global integration under the HSR environment

so that the travel time is reduced and the "same city effect" of the whole world is reached. Therefore, the upsurge of HSR construction has been launched worldwide. See Fig. 7.1.

7.1   The Continental Integration Under HSR Environment

At present, the four most mature countries in the development of HSR technology are Germany, France, Japan and China. French HSR trains (TGV) have become the world's most popular HSR technology with its excellent power control and intelligent protection systems. Among the countries with HSR lines already in operation, there are 6 countries using French technology for HSR train. Japan is the first country in the world to conduct research on HSR. Shinkansen technology is also known as the safest HSR technology in the world. Both CRH of Mainland China and the HSR of Taiwan adopt Japanese Shinkansen technology. China is the country with the latest start of HSR, but is also the fastest-growing country. China has designed five types of HSR trains such as CRH1, CRH2, CRH3, CRH5 and CRH380 through the introduction of HSR technology such as Japan, Germany and France. China has become the country with the longest mileage, fastest operation and largest line network in the world.

7.1.1   The European Village Under HSR Environment

Europe derives its name from the Greek mythology "Europa" (Greek: Ευρώπης). Europe is located in the northwestern part of the Eastern Hemisphere, bordering the Arctic Ocean to the north, the Atlantic Ocean to the west, and the Mediterranean

Sea and the Black Sea in the Atlantic Ocean to the south. Europe is bordered with Asia east to the Ural Mountains and the Ural River, southeast to the Caspian Sea, the Greater Caucasus Mountains and the Black Sea. Europe faces North America across the Atlantic Ocean, the Greenland Sea, the Danish Strait to the west, bordering the Arctic Sea to the north, and faces Africa across the Mediterranean Sea to the south (The dividing line is: Strait of Gibraltar). The northernmost end of Europe is Norwegian Nova Kok, the southernmost tip is the Maroki corner of Spain, and the westernmost end is the Portuguese Rocca. Europe, the world's second-largest continent, is only a little larger than Oceania, and it is called the Eurasian continent combined with Asia, and called Asian and European continents combined with Asia and Africa.

In 1994, a meeting of the European Commission was held in Germany, and it decided to implement the resolution to build and expand the pan-European transportation network. In 1998, the International Iron League began to organize further research on the European HSR network, and required the all-European HSR network to be formed in 2020. The European HSR network can be seen in Fig. 7.2.

7.1.2 The Asian Village Under HSR Environment

Asia (the ancient Greek language: Aσία; Latin: Asia), is the largest and most populous continent in seven continents. Most of Asia is located in the northern and eastern hemispheres. The dividing line between Asia and Africa is the Suez Canal. East of the Suez Canal is Asia. The dividing line between Asia and Europe is the Ural Mountains, the Ural River, the Caspian Sea, the Greater Caucasus Mountains, the Turkish Straits and the Black Sea. To the east of the Ural Mountains and the Greater Caucasus Mountains, the Caspian Sea and the Black Sea are Asia. The West is connected to Europe, forming the largest land mass on the continent which is called Eurasia.

The Trans-Asian Railway (TAR) is a unified freight rail network that runs through Eurasia. Representatives from 18 countries in Asia officially signed the Intergovernmental Agreement on the Asian Railway Network in Busan, South Korea on November 10, 2006. The plan for the Pan-Asian Railway Network, which has been planned for nearly 50 years, was finally implemented. According to the agreement's plan, in the near future, the four gold corridors of the "Steel Silk Road" will connect the two continents of Europe and Asia, and the crises-crossing trunk lines and feeder lines will weave a huge network of economic cooperation. According to the HSR line construction plan of China, Japan and other countries, the Asian ring, centered on China, east to Japan, west to Saudi Arabia, south to Malaysia, and north to Russia, will be realized by 2050. The Pan-Asian Line and Central Asia Line will form the Asian HSR network. However, at present, the countries concerned have to face the arduous task of unifying technical standards, coordinating customs, and quarantine and security inspection procedures, raising huge construction funds, and unifying the pace of construction. The Asian HSR network is shown in Fig. 7.3.

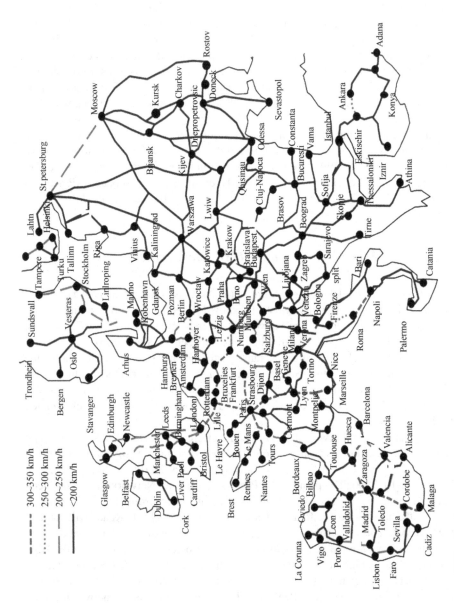

Fig. 7.2 European HSR network

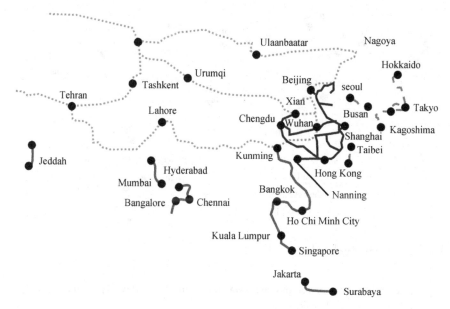

Fig. 7.3 Asian HSR network

(1)   China—Pakistan—China—Kyrgyzstan—Uzbekistan Rail. On April 20, 2015, China and Pakistan signed a framework agreement for the joint feasibility study of the No.1 rail trunk line (ML1) upgrade. The No.1 rail line runs from Karachi to Lahore and Islamabad to Peshawar, with a total length of 1,726 km. Figure 7.4 shows the China—Pakistan HSR line.

The idea of the Trans-Asian Railway is called the "Silk Road", which runs from Singapore, through Bangladesh, India, Pakistan and Iran to Istanbul, Turkey, and finally extends to Europe and Africa (Fig. 7.5).

(2)   Southeast Asia line. The Southeast Asia Line starts from Kunming of China, and is divided into three branches connecting Myanmar, Thailand and Cambodia. It is directly connected to Malaysia through Thailand. The Pan-Asian Line is a fast-moving ground transportation channel between China and Southeast Asia. It can improve the efficiency of tourism, economic and trade

Fig. 7.4 The China—Pakistan HSR line

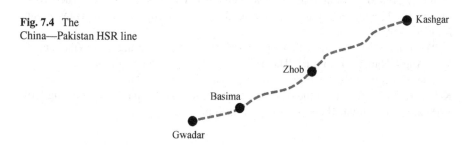

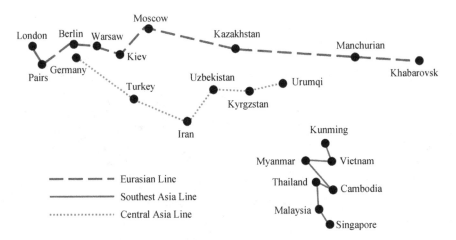

Fig. 7.5 The "Silk Road" of HSR

interactions between Southeast Asia, and promote the development of regional integration. See Fig. 7.6.

There are three programs through Kunming of China: First program is the east line program, from Singapore through Kuala Lumpur, Bangkok, Phnom Penh, Ho Chi Minh City, Hanoi to Kunming; Second program is the mid-line program from Singapore through Kuala Lumpur, Bangkok, Vientiane, Shangyong, Linyi, Xiangyun (Dali) to Kunming; Third program is the western route program from Singapore to Kuala Lumpur, Bangkok, Yangon, Ruili to Kunming. The quasi rail will replace the narrow rail.

The first plan: Singapore—Kuala Lumpur—Bangkok—Phnom Penh—Loc Nink of Ho Chi Minh—Hanoi—Lao Cai—Kunming. The total length of it is 5328 km. This line will also establish a branch line to connect Vientiane, the capital of Laos. The line is Vientiane—Tha Khet—Xinyi (Vung Ang Port). The total length of the branch line is 585 km. This line will also build a new road linking Vietnam, Laos and Cambodia with a total estimated cost of $1.8 billion.

The second plan: Singapore—Kuala Lumpur—Bangkok—Yangon—Kunming. The total length of it is 4559 km. In this line selection scheme, new sections will be built in Thailand, Myanmar and China to connect. The total length of the new road section is 1127 km, and the estimated total cost is $6 billion for the Trans-Asian Railway ASEAN channel.

The third plan (3A line): Singapore—Kuala Lumpur—Bangkok—Vientiane—VungAng—Hanoi—Kunming. The estimated cost of it is $1.1 billion.

The fourth plan (line 3B): Singapore—Kuala Lumpur—Bangkok—Vientiane—Kunming. The new line of the road network in Laos and China has a total length of 1300 km. The estimated cost of it is $5.7 billion.

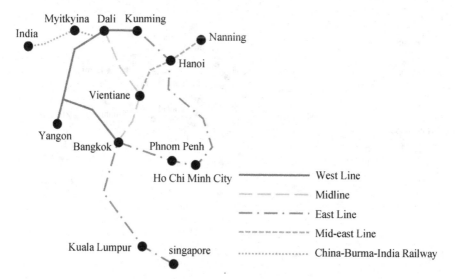

(a) Transasia railway planning drawing

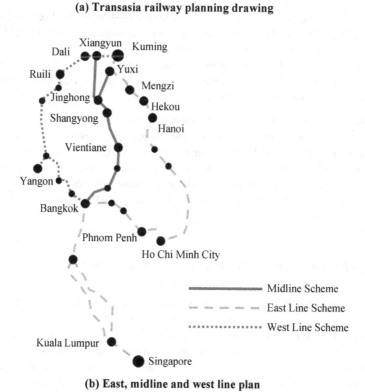

(b) East, midline and west line plan

Fig. 7.6 The HSR line of Southeast Asia

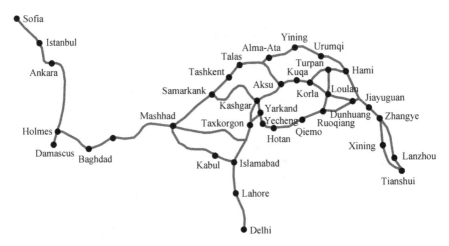

Fig. 7.7 HSR loop line of Asia pacific

The fifth plan (3C line): Singapore—Kuala Lumpur—Bangkok—Pacxe—Xavan-nakhet—Dong Ha—Hanoi—Kunming. The new line in this line has a total length of 616 km and an estimated cost of $1.1 billion.

The sixth plan (3D line): Singapore—Kuala Lumpur—Bangkok—Vientiane—Kunming. The line selection project is a combination of renovation and construction. The estimated cost of it is $1.1 billion.

(3) Central Asian line. Based on the China's four Xinxiang rails, this line is through the three cities of Harbin, Hohhot and Lhasa, respectively, to connect with Japan, Kazakhstan, India and other countries, and based on this, it continue to penetrate the Middle East, and finally reach Saudi Arabia. The line is conducive to cooperation between China and Middle East countries in energy, food, health, labor and many other aspects. It is the "Silk Road" for strengthening cooperation in the Asia–Pacific region. See Fig. 7.7.

7.1.3   The American Village Under HSR Environment

America is a combination of South America and North America. It is also the abbreviation of "Americas", also known as the new land. North America is located in the northern hemisphere. East Atlantic Ocean and West Pacific are north near to the Arctic Ocean, south to the Panama Canal and South America. In addition to including the Panama Canal to the north of North America, North America also includes the Caribbean Sea in the West Indies. South America is located in the southern part of the Western Hemisphere, with the Atlantic Ocean to the east, the Pacific Ocean to the west, the Caribbean Sea to the north, and facing the Antarctic across the Drake Strait.

Fig. 7.8 The HSR network
planning of North America

North America has not yet built the HSR in the real sense. The two largest countries in North America, Canada and the United States, have developed industries and the prosperous economies. Both two countries are now planning and they all have the conditions to build the HSR. Major cities in Canada are basically along the border of U.S., connecting Montreal from Quebec in the east, Ottawa in the capital, Toronto to the largest city, and Vancouver to the West coast of the Pacific Ocean. The U.S. has a vast territory and a population of more than 300 million, it is a highly developed economy. The total industrial and agricultural production scale ranks first in the world. Figure 7.8 shows the HSR network planning of North America.

South America is large in size and there are not many countries in it. Brazil is a well-developed industrial and agricultural country. It is a BRICS country. Other countries such as Argentina, Colombia, Venezuela, Peru, Chile and Ecuador are large countries. However, due to the financial, technical and political reasons, the HSR has not been built yet. Recently, Brazil has planned to build the HSR, especially in the capital city of Brasilia and Rio de Janeiro and São Paulo, as well as coastal cities. It is believed that in the near future, countries will soon build HSR according to the economic development, and will connect different countries' HSR lines in the ring coastline. South America will be linked to the U.S. and Canada by HSR from Panama, Costa Rica, Nicaragua, Honduras, Guatemala and Mexico. Figure 7.9 shows the HSR network planning of South America.

Fig. 7.9 The HSR network planning of South America

7.1.4 The African Village Under HSR Environment

Africa is located in the western part of the Eastern Hemisphere, South of Europe, West of Asia, the Indian Ocean to the southeast, and the Atlantic Ocean to the west, and longitudinal across the equator.

Africa is located in the westernmost part of the eastern hemisphere as the second largest continent in the world. Africa is bordered by the Atlantic Ocean in the west, Indian Ocean in the east, Europe across the Mediterranean Sea in the north, Asia in the northeast, and the equator runs through the Central Region. There are 57 countries and regions, of which Sudan has the largest area and Nigeria has the largest population. The rails are mostly narrow gauge, meter-rail, wide-track, and quasi-rail railways left by the original colonists with different specifications and different gauges. According to the characteristics of terrain and layout, most countries not only promote the standard gauges at home, but also connect together on the coastal quasi-rails. African countries will be linked by HSR when the economy sufficiently developed. Repairing the HSR along the coastline will reduce many of the cost. For example, Egypt has begun planning to build a 1000 km of HSR from Cairo to Aswan.

HSR is a high-input and high-cost project. At present, the routes with better comprehensive benefits of HSR are widely distributed among urban clusters with high population density and economic development. The large cities in Africa have experienced rapid population expansion in recent decades, but the degree of economic development has not been correspondingly improved. It is difficult to support the cost of HSR operations in the short term. Figure 7.10 shows the HSR network planning of Africa.

Fig. 7.10 The HSR network planning of Africa

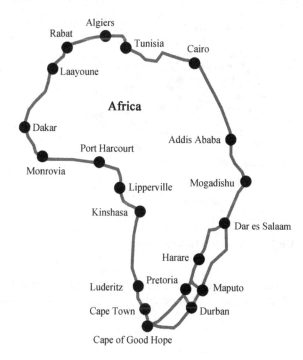

7.2 The Regional Integration Under HSR Environment

As a safe, reliable, fast and comfortable, large capacity, low carbon emission and environmentally friendly transportation mode, HSR has become an important trend in the development of the world transportation industry. The HSR planning route of world can be seen in Fig. 7.11.

7.2.1 The Asia-Europe Integration Under HSR Environment

The HSR has created a new growth point for urban development, promoted the integration of central cities and satellite towns, enhanced the radiation effect of central cities on surrounding cities, and strengthened the "same city effect" of neighboring large cities. In particular, with the development of HSR in various countries around the world, in the foreseeable future, high-speed rail will become the main ground transportation mode connecting countries in various regions, and will open the "global village" era together with air routes. Traveling around the world by HSR will become tourism fashion. Therefore, with the continuous improvement of HSR technology, the high-speed rail between countries and regions around the world is like a bus, with many shifts and short intervals, without the congestion of freeway and the delay of air plane. The travel time is reduced and the "same city effect" of the

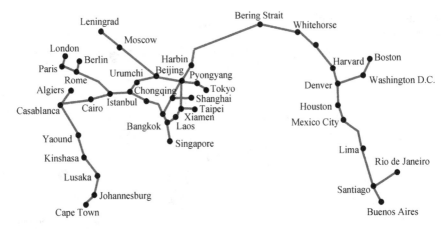

Fig. 7.11 The HSR planning route of world

whole world is reached. Therefore, the world has set off the upsurge of construction of HSR. Figure 7.12 shows the Asia-Europe HSR network.

Fig. 7.12 The Asia-Europe HSR network

7.2.2 The Africa-Europe Integration Under HSR Environment

The most concentrated HSR in Europe is in the western developed countries in the west. Apart from the fact that these countries have basically built a HSR network, they will consider the links with other countries. At the same time Europe will consider the three Nordic countries, HSR links Copenhagen (Denmark), Helsinki (Finland), Oslo (Norway). In addition, countries with better conditions, such as Poland, Romania, Bosnia and Herzegovina, are also preparing to build HSR. After the construction of HSR in the African the Spanish HSR can be considered to connect with Africa through the Gibraltar Strait Undersea Tunnel. Figure 7.13 shows the Europe-Africa HSR network.

Fig. 7.13 The Europe-Africa HSR network

7.2.3   The Europe-Asia-Africa Integration Under HSR Environment

The HSR is a revolution of road traffic and its main advantage is fast and safe. Countries with economic strength in the world are all planning to build the HSR, and many countries are jointly planning to build a transnational HSR network. Among them, 17 countries around China and China itself are negotiating the construction of the Asia-Europe Railway and the Trans-Asian Railway. Some routes of the Asia-Europe Railway and the Trans-Asian Railway are already under construction. Figure 7.14 shows the world HSR network planning.

The HSR network planning is mainly composed of three main lines: Pan-Asian Line, Asia-Europe Line and Transoceanic China-US Line. The details are as follows:

① The Pan-Asian line mainly connects to China and Southeast Asia such as Vietnam, Laos, Singapore and Bangkok.

② The Eurasian line mainly connects China, Russia, Britain, France and other Asian and European regions.

③ The China-US Transoceanic Line departs from Harbin, China, passes through the eastern part of Russia, crosses the Bering Strait, and travels through Canada to the United States.

According to China's HSR planning, in the near future, the world will form a HSR network as the main road transport channel, entering the "global village" era.

7.3   The Global Integration Under HSR Environment

In view of the considerable economic benefits and immeasurable political influence of high-speed rail, many countries in the world have invested in the construction of HSR. More than 20 countries including Germany, France, Italy, Spain, the United Kingdom, Japan, China, South Korea, the United States, and Brazil have opened

Fig. 7.14   The world HSR network planning

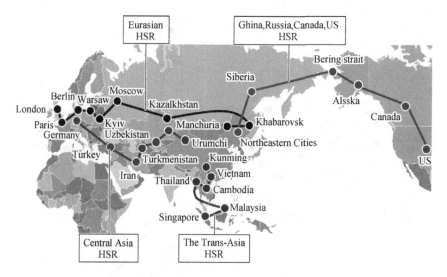

Fig. 7.15 The global HSR integration

HSR lines, and two countries, Russia and India are planning and preparing for the construction of HSR. Figure 7.15 shows the Global HSR integration.

① Eurasian HSR line: It starts from London, passes through Paris, Berlin, Warsaw, Kiev, after Moscow, and is divided into two routes, one into Kazakhstan, the other pointing to the Far East of Khabarovsk, and then enters Manzhouli, China.

② Central Asian HSR line: The starting point is Urumqi, this line runs through Kazakhstan, Uzbekistan, Turkmenistan, Iran, Turkey and other countries, and finally arrives in Germany.

③ Pan-Asian HSR line: It starts from Kunming, passes through Vietnam, Cambodia, Thailand, Malaysia, and finally arrives in Singapore. Process: The China-Myanmar rail tunnel starts in June.

④ China-Russian and Canadian-American HSR line: It starts from the northeast of China and travels north through Siberia to the Bering Strait. It builds a tunnel across the Pacific Ocean to Alaska, then from Alaska to Canada, and finally to the U.S.

7.3.1 Asia Pacific Loop Restarts "Silk Road"

According to the HSR construction plan of China, Japan and other countries, the Asian ring centered on China, east to Japan, west to Saudi Arabia, south to Malaysia, and north to Russia, the Asian HSR network formed by the Pan-Asian and Central Asian lines will be realized by 2050. See Fig. 7.16.

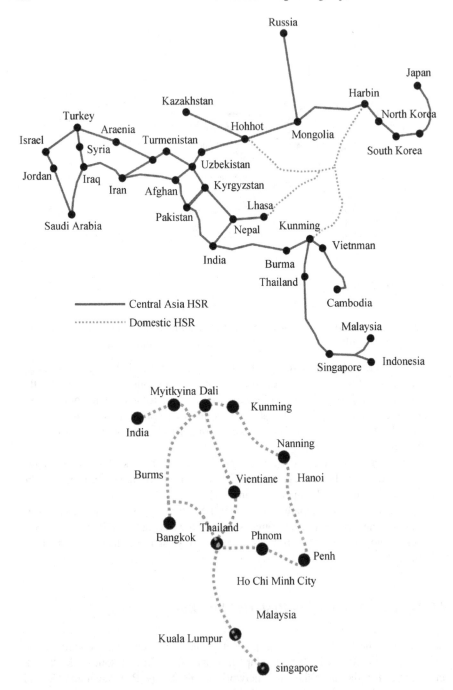

Fig. 7.16 The schematic of HSR line of Pan-Asian

7.3.2 Trans European Network Promotes European-African Integration

European HSR project "Trans European Network" is dedicated to building fast inter-city trains through Germany, France, Spain, Italy, the United Kingdom and other countries, shortening travel time between EU cities and promoting EU economic development and achieving European integration. The current HSR network in Europe is mainly centered on France, and it is extended to the internal railway network of other European countries through the northern line, the Atlantic line, the southeast line and the Mediterranean line to form a European intercity express train network. Among them, the northern line is the main channel for the British, Dutch and Germanic TGV to enter France. France goes straight to Spain through the Atlantic line and the southeast line connects France with Switzerland and Italy.

7.3.3 Eurasian Green Line Integrates "Pacific Rim Economic Circle"

The Eurasian HSR project starts from London and passes through Paris, Berlin, Warsaw, Kiev, and after Moscow, it is divided into two routes, one enters Kazakhstan, the other leads to Khabarovsk which is far away from the Far East, and finally enters China. The construction of the Eurasian HSR will accelerate the passenger and cargo transportation between the EU, Russia and China, weaken the barriers to regional cooperation caused by space constraints, and promote the integration of the "Pacific Rim Economic Circle". The traffic artery that runs through Eurasia will recombine the production factors of the countries along the route. China can become the driving force of the east, the EU is the driving force for the west, Russia is the driving force for the north, India is the driving force for the south, and progress in all directions will converge in the Middle East. The schematic of Eurasian HSR line can be seen in Fig. 7.17.

7.3.4 China-US Transoceanic Line Promotes the Formation of the "Global Village"

The connection between China's HSR and the US's HSR can be achieved mainly through two paths. One is entering Europe through China's Europe-Asia line, and then entering the Americas through Europe. However, in view of the science and technology in short time, it is difficult to realize HSR lines across the Atlantic hence, the above line cannot be implemented. The other one is that China's HSR may pass through the northeastern part of Russia, across the Bering Strait, into Canada, and finally into the U.S. This route is mainly from Beijing in China, through the Harbin

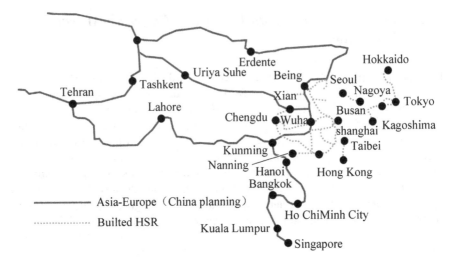

Fig. 7.17 The schematic of Eurasian HSR line

in the northeast to Russia's Yakutsk, through the eastern part of Russia to the Bering Strait, through the construction of the underwater tunnel into the Canadian western city of Fairbanks, along Whitehorse Fort McMurray and Edmonton, across Canada into Harvard, the northern city of the United States, and connected directly to Denver, the central city of the United States, through Harvard. The second planning route needs to overcome two technical problems along the way: on the one hand, it needs to cross the severe cold region of Russia in the east, and the HSR needs to overcome the safe operation under the cold and snowy conditions. Currently, it is basically solved successfully; On the other hand, crossing the Bering Strait, building a submarine tunnel is another technical problem that needs to be overcome. The technology is still in the process of tackling key problems. See Fig. 7.18.

China intends to cooperate with Russia, Canada and the United States to build a HSR line stretching 10,000 km that spans the Bering Strait and connects the two

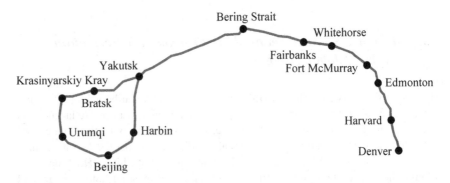

Fig. 7.18 The China-US HSR planning route

continents of Asia and the United States. The planned route runs north from the northeast, arrives at the Bering Strait via Siberia, crosses the Pacific Ocean by tunneling, arrives in Alaska, then goes to Canada, and finally arrives in the United States.

7.4 The Global Village Under SSR Environment

With the maturity of HSR technology, HSR will gradually replace the traditional railway lines and become a ground transportation method that competes with aviation. The HSR can not only shorten the travel time between regions, accelerate regional economic cooperation, but also have a strategic significance for the future world political situation. The planning and investment of the HSR is an important way to achieve integration with the world at an early date. Therefore, as a normalized vehicle that satisfies people's daily travel, the HSR in China has begun to transform into internationalization and globalization. Today, the HSR in China represents not only a means of transportation, but also the level of HSR and the spirit of the times, is a display of comprehensive strength.

By 2020, the total mileage of the global HSR will reach 42,000 km and the world will enter the "high-speed rail" era. With the integration of HSR in Asia, Europe, America and Africa, we can travel the world on HSR. China is standing at the forefront of the world's rails. China cooperates with the world to cope with global challenges, plan for the future, and create a better life for mankind.

7.5 Summary

In 1964, the world's first HSR was opened in Japan, and the first round of "high-speed rail heat" was launched worldwide. However, due to the technical problems, HSR has not been vigorously developed, and the operating speed is lower than 300 km/h. In 1995, French HSR technology became the technical standard for all-European HSR trains and was exported overseas. The second round of "high-speed rail heat" was launched worldwide. However, due to the economic downturn in the world, especially the limited economic capacity of developing countries, HSR was only built and operated in large and economically developed countries. In 2008, with the development of China's HSR, the world set off a third round of "high-speed rail heat". The HSR technology was developing in an advanced, mature, economic, applicable and reliable direction. In 2012, HSR was built and operated in developing counties because of the reduction of construction costs and the economic development of developing countries.

The development of the HSR has a major impact on national and regional development strategies. At present, most countries in the world have begun to build a transnational HSR, as soon as possible to achieve the rapid road access between countries and regions, eliminate the impact of geographical restrictions between countries, and accelerate exchanges and cooperation between regions and regions. Considering the overall situation of national development, the development of the HSR has far-reaching strategic influences.

Strategically speaking, it helps to protect the national security. Considering the overall development of the world, The HSR has a profound impact on the world's political economy. The HSR can promote world integration and realize the "global village".

References

Akgungor AP, Demirel A (2007) Evaluation of Ankara—Istanbul high-speed train project. Transport 22(1)

Chen Q (2013a) Super high-speed rail leads the new speed in the future. Mod Ind Econ Informationization 17(13):80–81. (In Chinese)

Chen X (2013b) Aerodynamic simulation analysis of evacuated tube maglev trains. Southwest Jiaotong University. (In Chinese)

Chen G (2017) China's high-speed rail has created many world firsts. Traffic Transp (02):44. (In Chinese)

Fang Q (2003) Main technical features of high-speed rail and high-speed EMU. Electr Drive Locomotives (05):5–9. (In Chinese)

Gu J (2016) Spike magnetic levitation vehicle—super high-speed rail. China Policy Review. (In Chinese)

Hu S, Xu Y (1993) Technical and economic analysis of tr magnetic suspension high-speed rail system. J Beijing Jiaotong Univ (3):233–237. (In Chinese)

Hu S, Zhang J (1993) German magnetic levitation high-speed rail system. World Railway (2):17–19. (In Chinese)

Jie H (2016) Accessibility of chinese provincial capitals based on the presence of high-speed rail. Jiangxi Noraml University. (In Chinese)

Kaliankovich V, New evolutionary model of the transport and communication interaction between Russia, China and Europe. Lanzhou Jiaotong University. (In Chinese)

Ken N, Zhou X (2017) Research and development concerning superconducting Maglev and research on applying its technology to the conventional railways. Foreign Rolling Stock 54(2):25–28. (In Chinese)

Li Y (2013) The world has entered the era of high-speed rail. Traffic Transp (01):17–18. (In Chinese)

Li P, Wang T, Song Y (2013) Realization and technology research of 3D simulation system of high-speed train based on virtools. Chinese Railways (07):40–42+47. (In Chinese)

Liang X, Qi M, Tan K (2107) Study on business modes of Pan-Europe high-speed rail express transportation. Railway Transp Econ (02):79–84. (In Chinese)

Lin G, Lian J (1998) Development of Japan Maglev high-speed rail and technical and system characteristics of Yamanashi Test Line. Electr Drive Locomotives (4):5–8. (In Chinese)

Lu C (2015) Highlights of China high-speed rail. Sci Technol Rev (18):13–19. (In Chinese)

Lv Z (2105) The research on the sustainable competitive advantage of china's high-speed rail. Beijing Jiaotong University. (In Chinese)

Mnich P, Wang B (2001) Current Status and comparison of German and Japanese Maglev high-speed rail systems. Converter Technol Electr Traction (6):1–8. (In Chinese)

Peng Q, Li J, Yang Y et al (2016) Influences of high-speed rail construction on railway transportation of China. J Southwest Jiaotong Univ (03):525–533. (In Chinese)

© Southwest Jiaotong University Press 2023
Q. Hu and S. Qu, *A Brief History of High-Speed Rail*,
https://doi.org/10.1007/978-981-19-3635-7

Sheng R (2016) Research on grounding protection of medium and low speed Maglev traffic traction power supply system. J Railway Eng Soc 33(10). (In Chinese)

Sheng Y, Hu S (2002) Comparative study on passing capacity of high-speed rail and high-speed Maglev railway. Technol Econ (9):39–40. (In Chinese)

Sun F, Wang D, Niu Y (2017) Competition patterns of high-speed rail versus freeways and aviation. Geogr Res (01):171–187. (In Chinese)

(2013) "Super high-speed rail": the carriage is flying in the vacuum pipe. Mech Eng. (In Chinese)

(2016) Super high-speed rail: advance like a cannonball. Popular Sci. (In Chinese)

Tang Z, Sun J, Wu L (2014) Aerodynamics analysis and optimization design of ETT. Mach Tool Hydraul (21). (In Chinese)

Thompson C (2015) The next pipe dream. Smithsonian 46(4):17–23

Tian T, Li T, Li G et al (2016) Investigation of dynamic response model of high-speed rail tunnel lining structure with vault cavity. Chin J Undergr Space Eng (S2):669–677. (In Chinese)

Wang F (2012) Study on the influence of China's high-speed rail on regional economic development. Jilin University. (In Chinese)

Wang Y (2016) Review and prospection of China's high-speed rail. Railway Econ Res (01):6–11. (In Chinese)

Wang S, Wang J (1999) High temperature superconducting magnetic levitation. Cryogenics Supercond (4):8–12. (In Chinese)

Xing P (2017) The formation and development of China's high-speed rail network. Contemp Econ (03):28–30. (In Chinese)

Xu K (2014) The investigation of the humanistic design and use of the high-speed passenger railway station. Fudan University. (In Chinese)

Yang H (2016) Super high-speed rail: cross-border performance of high-speed spacecraft[N]. China Aerospace News, 2016-05-21:004. (In Chinese)

Yin P (2012) High-speed rail (HSR) and establishment of new pattern of regional tourism: a case study of the high-speed rail between Zhengzhou and Xi'an. Tourism Tribune (12):47–53. (In Chinese)

Zhang P (2016) Can "super high-speed rail" start smoothly?. People's Wkly. (In Chinese)

Zhao Y, Shang Y (2017) Study on dynamic characteristics and cumulative deformation law of subgrade under dynamic load of high-speed train. Railway Stand Des (07):56–61. (In Chinese)

Zhao H, Dong B, Yu Z (2013) Study on characteristics of traveling selection between high-speed rail and airline passengers. Railway Transp Econ (11):32–36. (In Chinese)

Zhao Y, Li X, Wei G (2015) Study on Influence of high-speed rail on regional economic system. Railway Transp Econ (03):7–13. (In Chinese)

Zhou X, Zhang Y, Yao Y (2008) Numerical simulation of air resistance of high-speed trains in vacuum pipes. Sci Technol Eng 8(6). (In Chinese)

Zhenmin Z (2002) 21st century superconducting maglev train. Railway Signal Commun 38(6):39–41 (In Chinese)

Printed in the United States
by Baker & Taylor Publisher Services

Printed in the United States
by Baker & Taylor Publisher Services